D0151699

Real Infinite Series

Library of Congress Control Number 2005937268
ISBN 0-88385-745-6
Printed in the United States of America
Current Printing (last digit):
10 9 8 7 6 5 4 3 2 1

Real Infinite Series

Daniel D. Bonar
Denison University

and

Michael Khoury, Jr.
Ohio State University

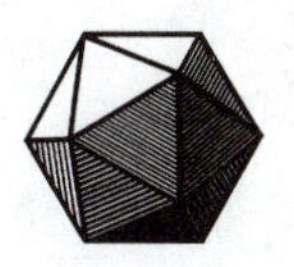

Published and Distributed by
The Mathematical Association of America

CLASSROOM RESOURCE MATERIALS

Classroom Resource Materials is intended to provide supplementary classroom material for students—laboratory exercises, projects, historical information, textbooks with unusual approaches for presenting mathematical ideas, career information, etc.

101 Careers in Mathematics, 2nd edition edited by Andrew Sterrett

Archimedes: What Did He Do Besides Cry Eureka?, Sherman Stein

Calculus Mysteries and Thrillers, R. Grant Woods

Combinatorics: A Problem Oriented Approach, Daniel A. Marcus

Conjecture and Proof, Miklós Laczkovich

A Course in Mathematical Modeling, Douglas Mooney and Randall Swift

Cryptological Mathematics, Robert Edward Lewand

Elementary Mathematical Models, Dan Kalman

Environmental Mathematics in the Classroom, edited by B. A. Fusaro and P. C. Kenschaft

Essentials of Mathematics, Margie Hale

Exploratory Examples for Real Analysis, Joanne E. Snow and Kirk E. Weller

Fourier Series, Rajendra Bhatia

Geometry From Africa: Mathematical and Educational Explorations, Paulus Gerdes

Historical Modules for the Teaching and Learning of Mathematics (CD), edited by Victor Katz and Karen Dee Michalowicz

Identification Numbers and Check Digit Schemes, Joseph Kirtland

Interdisciplinary Lively Application Projects, edited by Chris Arney

Inverse Problems: Activities for Undergraduates, Charles W. Groetsch

Laboratory Experiences in Group Theory, Ellen Maycock Parker

Learn from the Masters, Frank Swetz, John Fauvel, Otto Bekken, Bengt Johansson, and Victor Katz

Mathematical Connections: A Companion for Teachers and Others, Al Cuoco

Mathematical Evolutions, edited by Abe Shenitzer and John Stillwell

Mathematical Modeling in the Environment, Charles Hadlock

Mathematics for Business Decisions Part 1: Probability and Simulation (electronic textbook), Richard B. Thompson and Christopher G. Lamoureux

Mathematics for Business Decisions Part 2: Calculus and Optimization (electronic textbook), Richard B. Thompson and Christopher G. Lamoureux

Ordinary Differential Equations: A Brief Eclectic Tour, David A. Sánchez

Oval Track and Other Permutation Puzzles, John O. Kiltinen

A Primer of Abstract Mathematics, Robert B. Ash

Proofs Without Words, Roger B. Nelsen

Proofs Without Words II, Roger B. Nelsen

A Radical Approach to Real Analysis, David M. Bressoud

Real Infinite Series, Daniel D. Bonar and Michael Khoury, Jr.

She Does Math!, edited by Marla Parker

Solve This: Math Activities for Students and Clubs, James S. Tanton

Student Manual for Mathematics for Business Decisions Part 1: Probability and Simulation, David Williamson, Marilou Mendel, Julie Tarr, and Deborah Yoklic

Student Manual for Mathematics for Business Decisions Part 2: Calculus and Optimization, David Williamson, Marilou Mendel, Julie Tarr, and Deborah Yoklic

Teaching Statistics Using Baseball, Jim Albert

Topology Now!, Robert Messer and Philip Straffin

Understanding our Quantitative World, Janet Andersen and Todd Swanson

Writing Projects for Mathematics Courses: Crushed Clowns, Cars, and Coffee to Go, Annalisa Crannell, Gavin LaRose, Thomas Ratliff, Elyn Rykken

MAA Service Center
P.O. Box 91112
Washington, DC 20090-1112
1-800-331-1MAA FAX: 1-301-206-9789

Preface

The theory of infinite series is an especially interesting mathematical construct due to its wealth of surprising results. In its most basic setting, infinite series is the vehicle we use to extend finite addition to "infinite addition." The material we present deals almost exclusively with real infinite series whose terms are constants. It is possible to spend an entire professional career studying infinite series, and there is no shortage of books written for such specialists. However, the authors of this book expect no such prerequisites from our readers. We believe that the general mathematician can and should understand infinite series beyond the Calculus II level; tragically, most existing presentations are either very basic or exclusively for the specialist. Between these covers, the reader will find discussions that are deep enough and rich enough to be useful to anyone involved in mathematics or the sciences, but that are accessible to all. Indeed, even those who are not mathematicians will be able to appreciate infinite series on an aesthetic level. This book is designed to be a resource on infinite series that may prove helpful to 1) teachers of high school calculus, 2) teachers of college level calculus, 3) undergraduate math majors who have had a taste of infinite series by way of one or two chapters in a beginning calculus text and need to review that material while learning much more, or 4) students who have studied infinite series in advanced calculus or in elementary real analysis and have a thirst for more information on the topic. This book may also serve as a resource or supplemental text for a course on real infinite series for either advanced undergraduates or beginning graduate students. It may be of value in connection with a special topics course, or as one of several texts for a seminar on infinite series.

Chapter 1 is devoted to a review of the basic definitions and theory of infinite series as found about midway through a thousand page beginning calculus text. This presentation is self-contained, so the reader who has not taken calculus or whose recollection of the idea of "limits" is shaky should in no way feel discouraged. The various tests for convergence/divergence such as the Comparison Test, the Limit Comparison Test, the Ratio Test, the Root Test, the Integral Test, etc. are presented in this chapter. While the mathematical content does not differ from that in many standard texts, the presentation is designed to give some understanding of the intuitive notion of an infinite series and to reconcile the definitions with an intuitive perspective.

Most students and mathematicians know basic tests such as the Ratio and Root Tests—they are taught in most advanced calculus classes and are easy to apply and to remember. However, several more powerful tests that are only slightly more technical are largely unknown. This is particularly unfortunate because many quite natural series elude analysis by these means; we believe that this can give the false impression that the study of infinite series has very limited applications for the nonspecialist. Chapter 2 is given over to these more delicate and sophisticated tests for convergence/divergence which one can use if the basic tests of Chapter 1 are insufficient; the reader will see that the greatly increased power comes with only small costs in complexity. Semi-well-known tests such as Raabe's Test, Kummer's Test, Cauchy's Condensation Test, Abel's Test, and Dirichlet's Test are discussed, and examples are offered to illustrate their worth, as well as tests such as Bertrand's Test, whose obscurity belies their elegance and power.

The harmonic series and the alternating harmonic series have intrigued mathematicians for centuries. Chapter 3 is largely devoted to facts concerning these two series and similar results for closely related series. Even exploring the origin of the name "harmonic" leads to interesting discoveries. While it is generally *known* that the harmonic series diverges while the alternating harmonic series converges, we believe that this exposition makes these statements more tangible, and we include several proofs of different types. This chapter deals with examples of rearranging series and some clever techniques for summing special series exactly. The exploration of the harmonic and alternating harmonic series gives some insight into more general series.

In some sense we see Chapter 4 being the heart of the book. We work through 107 gems of mathematics relating in one way or another to infinite series. One purpose of this chapter is to demonstrate that the theory of infinite series is full of surprising and intriguing results. These results have many forms: they may show the exact sum of a series, they may exhibit a neat proof of a common sense result, or they may present a particularly slick problem-solving technique. The gems presented here could be used as nontraditional examples in a classroom setting, or simply for the reader's own enjoyment, which is justification enough. Some of these gems even allow us to push the material of the preceding chapters even further, giving the reader a taste of extensions such as multiple series and power series.

The next chapter is dedicated to the annual Putnam Mathematics Competition, which has been in existence since 1938. In Chapter 5 we have reproduced nearly all the infinite series problems offered on those examinations. The reader who is so inclined may choose to take this as a challenge; however, solutions to the problems, as found in the literature, are also provided, in many cases accompanied by alternate solutions or commentary provided by the authors. Students of mathematics can learn a nice assortment of problem solving techniques by studying these problems and their solutions.

The final official chapter is offered to the reader as a parting gift, and is fittingly called "Final Diversions." Here we explore the lighter side of infinite series. We provide three puzzles, including the classic pebbling problem. Also we have a selection of visual materials, some presenting a proof of an important result in a pictorial way, and others providing a new way of thinking about infinite series. We hope the reader will also enjoy reading and considering some fallacies, presented for amusement but also to refine understanding; these are flawed arguments that might seem convincing if passed over too quickly.

Finally, we offer three appendices. Appendix A is a listing of 101 statements that the reader may identify as either true or false. True-False statements have an educational value in clarifying concepts that may seem confusing or counterintuitive. Each true-false statement is answered and explained. In Appendix B, we reprint an article by David E. Kullman about the harmonic series, which will be of special interest after reading our Chapter 3. Appendix C is an extensive (but certainly not exhaustive) collection of references. We provide a list of books worth reading, as well as a list of articles on infinite series found in three MAA publications: *The College Mathematics Journal*[1], *Mathematics Magazine*, and *The American Mathematical Monthly*. We will consistently abbreviate these titles as *CMJ*, *MM*, and *AMM*, respectively. These articles can be easily and quickly accessed via JSTOR.

We close the preface with an invitation to read and an invitation to enjoy. The subject of infinite series is a deep, intriguing, and useful one, and we hope the reader will also find it so. We believe this book will be a valuable introduction for new readers, as well as a companion and reference for all. Above all, we believe you will enjoy this book as much as we have enjoyed putting it together.

Acknowledgements

We are pleased to acknowledge those individuals who assisted in bringing this book to its final form. Andrew Sterrett at Denison University initially encouraged the first author to write a book on infinite series. What followed from that encouragement was a summer research project and a one semester directed study on infinite series involving both authors who then worked jointly to write this book.

Ed Merkes headed up the review process and provided us with valuable feedback. The anonymous reviewers of this work made suggestions that improved our early drafts. Zaven Karian generously offered guidance throughout the entire writing effort. Others who assisted us in developing this work include Professor Frank Carroll, Professor William Dunham, Elizabeth Ehret, Professor Henry Ricardo, Tony Silveira, and Helen Viles. We express our appreciation and thanks to these individuals. We extend thanks to the authors of the many articles and books listed in Appendix C. Their scholarship directly or indirectly contributed to the final form of this book. The visuals in section 6.2 were created by the individual(s) identified with each picture or diagram and we offer our thanks to them. Don Albers, Elaine Pedreira, Beverly Ruedi and the Publications Department at the Mathematical Association of America played a vital role in producing an attractive finished product and we convey our thanks for their role in having this book "see the light of day." The first author offers a special thank you to Denison University for a Robert C. Good Fellowship (a one semester leave) to work on infinite series.

And finally, we are especially grateful for the love and support of our families.

[1]In the first years of its publication, *The College Mathematics Journal* was called *The Two Year College Mathematics Journal*. We will consider these to be essentially the same publication and refer to both by the same symbol, trusting that the volume number will always make it clear which is being referred to.

Contents

1

Introduction to Infinite Series

The purpose of this introductory chapter is two-fold. First, we wish to lay out the basic definitions and theorems in the theory of infinite series, so that we will have groundwork established for our later, less well-known results. There is nothing in this chapter that the reader cannot find in a good undergraduate calculus textbook, so we neither intend nor claim to give the best or most thorough treatment of these basics. While we suppose that the reader is already familiar with infinite series, we do not require it. In Section 1.1, we leave for the reader the proofs of the *most* basic propositions about sequences and series; those who have no experience with infinite series should actually undertake these proofs. Detailed proofs can be found in many sources. In particular, we highly recommend that the reader who is not familiar with lim sup and lim inf consult another source, since we do little more here than give the definitions. A list of recommended sources is listed in the appendices.

At the same time, the authors of this book are aware that infinite series can be extremely confusing when they are first learned. This confusion leaves many scientists and general math students suspicious and fearful of infinite series, when in fact series add a valuable tool to a person's mathematics repertoire. In this chapter, when the information content is probably known to the reader, there is a focus on motivating the basic theorems and addressing some of the common questions and concerns.

1.1 Definitions

1.1.1 Infinite Sequences

Definition 1.1 *A (real) infinite sequence is a function from the nonnegative integers to the reals:*

$$a : n \mapsto a_n.$$

In practice we think of a sequence as an ordered infinite list $a_0, a_1, a_2, \ldots, a_n, \ldots$, and it may also be written $\{a_n\}_{n=0}^{\infty}$ or just $\{a_n\}$. The variable n is called the *index* and is a dummy variable; although sequences will generally be indexed from 0 to ∞, we allow sequences that begin at 1, or 13, or any other positive integer.

Definition 1.2 *A sequence $\{a_n\}$ is said to converge to a real number A if, for each $\epsilon > 0$, there exists an integer N such that for all $n > N$, $|a_n - A| < \epsilon$. In this case we write*

$$\lim_{n\to\infty} a_n = A$$

and say that $\{a_n\}$ is convergent with limit A or simply that the limit is A.

Of course, the definition would not change substantially if we insisted that N be a positive integer or if we relaxed the requirement that N be an integer altogether, or if we changed the inequality $n > N$ to a nonstrict inequality. Though we will try to be consistent in our usage here, the reader will encounter any number of minor variations on this definition in the literature. All of them, however, are equivalent to one another, and in particular to this definition.

Definition 1.3 *A sequence $\{a_n\}$ is said to diverge or be divergent if it does not converge to any real number A in the sense of the preceding definition.*

Definition 1.4 *A sequence $\{a_n\}$ is said to diverge to ∞ if, for any $M > 0$, there exists an integer N such that for all $n > N$, $a_n > M$. In this case we may write*

$$\lim_{n\to\infty} a_n = \infty.$$

Definition 1.5 *A sequence $\{a_n\}$ is said to diverge to $-\infty$ if, for any $M > 0$, there exists an integer N such that for all $n > N$, $a_n < -M$. In this case we may write*

$$\lim_{n\to\infty} a_n = -\infty.$$

We remark that, where we say a series diverges to ∞, some authors say converges to infinity. It is a matter of perspective, of whether we view ∞ as a particular number to which a sequence can converge or whether we consider this behavior a particular way in which a sequence can fail to converge to a finite number. The authors of this book take the second of these perspectives and reject the first. Presumably the first view is motivated by the similar lim notation that is used in either case. However, the necessity of a wholly different definition in the finite and infinite cases distinguishes them as fundamentally different types of behavior. There would be no question of making a definition based on making the meaningless difference $|a_n - \infty|$ small. While we acknowledge that the literature is inconsistent on this point, we will use "convergence" strictly in the sense of convergence to a finite limit. Another common convention, which we may use, is to speak of a sequence that diverges to one of the infinities as diverging properly. All of these remarks, incidentally, apply equally well to convergence and divergence terminology in the context of infinite series.

Definition 1.6 *A sequence that neither converges nor diverges to $\pm\infty$ is said to oscillate or diverge by oscillation.*

Example 1.7

$$a_n = \frac{(-1)^n n}{n+1}$$

oscillates with

$$\lim_{n\to\infty} a_{2n} = 1$$

and

$$\lim_{n\to\infty} a_{2n+1} = -1.$$

Proposition 1.8 (Basic Limit Properties) *The following statements about limits and sequences hold.*

1. *A convergent sequence has a unique (finite) limit.*
2. *A convergent sequence cannot also diverge properly.*
3. *If $\lim_{n\to\infty} a_n = A$, then $\lim_{n\to\infty}(ka_n) = kA$. (k a constant)*
4. *If $\lim_{n\to\infty} a_n = A$ and $\lim_{n\to\infty} b_n = B$, then $\lim_{n\to\infty}(a_n + b_n) = A + B$.*
5. *If $\lim_{n\to\infty} a_n = A$ and $\lim_{n\to\infty} b_n = B$, then $\lim_{n\to\infty}(a_n b_n) = AB$.*
6. *If $\lim_{n\to\infty} a_n = A$ and $\lim_{n\to\infty} b_n = B$, where b_n, $B \neq 0$, then $\lim_{n\to\infty}(a_n/b_n) = A/B$.*

Remark. The first two components of this proposition taken together guarantee that $\lim_{n\to\infty} a_n$, whenever it is meaningful at all, is a well-defined object and can, in general, be manipulated algebraically. Analogues of the remaining statements remain true when A or B is infinite, but we will leave the details to the interested reader.

Definition 1.9 *A sequence $\{a_n\}_{n=1}^{\infty}$ is bounded if and only if there exists a number M such that $|a_n| \leq M$ for all $n \geq 1$.*

Definition 1.10 *A sequence $\{a_n\}_{n=1}^{\infty}$ is called monotone increasing (or nondecreasing) if and only if $a_{n+1} \geq a_n$ for all $n \geq 1$. The sequence is called monotone decreasing (or nonincreasing) if and only if $a_{n+1} \leq a_n$ for all $n \geq 1$. The sequence is called monotonic if and only if it is either monotone increasing or monotone decreasing.*

The theorem that follows is used frequently in working with sequences and series. Its proof is based on the Completeness Axiom for the set of real numbers. The Completeness Axiom is an assertion of the fact that the real line has no gaps or holes. More precisely the Completeness Axiom says that any nonempty set S of real numbers that is bounded above (i.e., $x \leq M$ for all $x \in S$ and some M) has a least upper bound. Also if S is bounded below (i.e., $x \geq m$ for all $x \in S$ and some m) then it has a greatest lower bound. For example, if $a_n = n/(n+1)$, $n = 1, 2, 3, \ldots$, then $\{a_n\}$ has many upper and lower bounds but 1 is the least upper bound and $\frac{1}{2}$ is the greatest lower bound.

Theorem 1.11 *Every bounded monotonic sequence is convergent.*

Proof. Assume $\{a_n\}$ is monotone increasing and bounded above. By the Completeness Axiom $S = \{a_n : n = 1, 2, \ldots\}$ has a least upper bound; call it L. Given $\epsilon > 0$, $L - \epsilon$ is not an upper bound, hence $a_N > L - \epsilon$ for some integer N. Since $\{a_n\}$ is monotone increasing we also have $a_n > L - \epsilon$ for all $n \geq N$. Therefore $L - \epsilon < a_n < L + \epsilon$ for all $n \geq N$ or equivalently $\lim_{n\to\infty} a_n = L$. A similar proof holds if $\{a_n\}$ is monotone decreasing.

Definition 1.12 *Formally, a subsequence of a sequence $\{a_n\}$ is any sequence of the form $\{b_m\}$, where $b_m = a_{n_m}$, and the n_m are integers with*

$$n_1 < n_2 < n_3 < \cdots .$$

Informally, a subsequence is formed by considering any infinite subcollection of the terms of a sequence without changing their order.

Example 1.13 *Consider the sequence* $\{1/n\}_{n=1}^{\infty} = 1, 1/2, 1/3, 1/4, \ldots$. *Then one subsequence of this sequence would be* $1, 1/4, 1/9, 1/16, \ldots, 1/n^2, \ldots$.

Definition 1.14 *A real number A is called a subsequential limit point of a sequence if the sequence has some subsequence that converges to A. In addition we call ∞ (resp. $-\infty$) a subsequential limit point of a sequence if the sequence has a subsequence that diverges to ∞ (resp. $-\infty$).*

At this point, we list and illustrate a collection of additional facts concerning infinite sequences. All of these results are basic and can be proven easily, but awareness of them can make it much easier to work with and appreciate the theory of infinite sequences. We do not give the proofs here, but leave them to the reader; proofs can be found in typical analysis books.

Fact 1.15 *Changing a finite number of terms in a sequence has no effect on convergence, divergence, or the limit if it exists. For example, the sequences*

$$1, \frac{1}{2}, \frac{1}{3}, \frac{1}{4}, \frac{1}{5}, \frac{1}{6}, \ldots, \frac{1}{n}, \ldots,$$

$$2, 7, 5, \frac{1}{10}, \frac{1}{5}, \frac{1}{6}, \ldots, \frac{1}{n}, \ldots$$

both converge to 0.

Fact 1.16 *If all terms in a sequence are constant from some point on, the sequence converges to that constant. For example, the sequence*

$$1, 2, 3, 4, 5, 6, 7, 7, 7, 7, \ldots$$

converges to 7.

Fact 1.17 *Any subsequence of a convergent sequence converges, and its limit is the limit of the original sequence. For example,*

$$1, \frac{1}{2}, \frac{1}{3}, \frac{1}{4}, \frac{1}{5}, \frac{1}{6}, \ldots, \frac{1}{n}, \ldots$$

converges to 0 *and so also does*

$$\frac{1}{2}, \frac{1}{4}, \frac{1}{6}, \frac{1}{8}, \frac{1}{10}, \ldots, \frac{1}{2n}, \ldots.$$

Fact 1.18 *Any subsequence of a sequence that diverges to ∞ also diverges to ∞. For example,*

$$1, 2, 3, 4, 5, \ldots, n, \ldots$$

and

$$1, 8, 27, 64, 125, \ldots, n^3, \ldots.$$

both diverge to ∞.

Fact 1.19 *Any convergent sequence is bounded, but the converse fails. There are bounded sequences that do not converge. For example,*

$$\frac{1}{2}, \frac{2}{3}, \frac{3}{4}, \frac{4}{5}, \ldots, \frac{n}{n+1}, \ldots$$

is convergent and

$$\left|\frac{n}{n+1}\right| \leq 1.$$

On the other hand,

$$1, 2, 3, 1, 2, 3, 1, 2, 3, 1, 2, 3, \ldots$$

is bounded but not convergent.

Fact 1.20 *If $\{a_n\}$ converges to zero, and $\{b_n\}$ converges, then $\{a_n b_n\}$ converges to zero. As an example, consider $a_n = 1/n$, $b_n = n/n + 1$.*

Fact 1.21 *If $\{a_n\}$ is a sequence of nonzero numbers, then $|a_n| \to \infty$ if and only if $1/|a_n| \to 0$. Equivalently, $|a_n| \to 0$ if and only if $1/|a_n| \to \infty$.*

Fact 1.22 *If $a_n \le b_n$ for all n, and if $\lim_{n\to\infty} a_n$ and $\lim_{n\to\infty} b_n$ exist (whether finite or $\pm\infty$), then*

$$\lim_{n\to\infty} a_n \le \lim_{n\to\infty} b_n.$$

As one example, consider $a_n = 1/2^n$ and $b_n = 1/n$. As another, consider $a_n = n$, $b_n = n^2$.

Fact 1.23 *If two subsequences of a given sequence converge to distinct limits, then the sequence diverges. Consider for example*

$$1, 2, 3, 1, 2, 3, 1, 2, 3, 1, 2, 3, \ldots$$

which contains subsequences

$$1, 1, 1, 1, 1, \ldots$$

and

$$2, 2, 2, 2, 2, \ldots$$

convergent to limits 1 and 2.

Fact 1.24 *The sum, difference, product, and quotient of two divergent sequences need not diverge. For example, take $a_n = (-1)^{n+1}$, $b_n = (-1)^n$. Then $\{a_n + b_n\}$, $\{a_n b_n\}$, $\{a_n/b_n\}$ all converge. If we take $c_n = (-1)^{n+1}$, then $\{a_n - c_n\}$ converges.*

Fact 1.25 *If $\{a_n + b_n\}$ and $\{a_n - b_n\}$ both converge, say to L_1 and L_2, then $\{a_n\}$ converges to $(L_1 + L_2)/2$ and $\{b_n\}$ converges to $(L_1 - L_2)/2$.*

Fact 1.26 $\lim_{n\to\infty} a_n = 0$ *if and only if* $\lim_{n\to\infty} |a_n| = 0$.

Fact 1.27 *If $\lim_{n\to\infty} a_n = a$, then $\lim_{n\to\infty} |a_n| = |a|$. However, if $a \ne 0$, then the converse does not hold. An example in which the converse fails is*

$$a_n = \frac{(-1)^n 2n}{n+1}.$$

Here $\{|a_n|\}$ converges to 2, but $\{a_n\}$ does not converge either to 2 or to -2.

Fact 1.28 *If $a_n \le M$ for all n and $a_n \to a$, then $a \le M$. However, even if $a_n < M$ for all M, we may not have $a < M$. For example, if $a_n = 1 + 1/n \le 2 = M$, we have $\lim_{n\to\infty} a_n = 1 \le 2$. On the other hand, if $a_n = n/n + 1$, $a_n < 1$, but $\lim_{n\to\infty} a_n = 1$.*

Fact 1.29 *If $\{a_n\}$ is a decreasing sequence, then there are two possibilities:*

1. *$\{a_n\}$ is bounded below by M, in which case it converges to a number $L \ge M$,*
2. *$\{a_n\}$ is not bounded below, in which case it diverges to $-\infty$.*

The sequences $\{1 + 1/n\}$ *and* $\{-n\}$ *illustrate the two possibilities.*

Fact 1.30 *If* $\{a_n\}$ *is an increasing sequence, then there are two possibilities:*

1. $\{a_n\}$ *is bounded above by* M*, in which case it converges to a number* $L \leq M$*,*
2. $\{a_n\}$ *is not bounded above, in which case it diverges to* ∞*.*

Definition 1.31 *The limit superior (resp. limit inferior) of a sequence* $\{a_n\}$ *is the least upper bound (resp. greatest lower bound) of all the subsequential limit points for the sequence. We write this* $\limsup_{n\to\infty} a_n$ *(resp.* $\liminf_{n\to\infty} a_n$*).*

These concepts have an advantage over the more natural-looking concept of limit in the sense that the limit superior and limit inferior exist for any sequence (though one or both will be infinite if the sequence fails to be bounded), and so give information about sequences that are not sufficiently well-behaved for the ordinary limit to exist. Another convenient property that we will not be using and consequently will not prove is that, if the limit superior is in fact finite, it is actually the *maximum* of the subsequential limit points, a stronger property than merely being the least upper bound. Naturally an analagous statement holds for the limit inferior.

Proposition 1.32 (Basic Limit Superior/Inferior Properties) *The following facts about limits superior and limits inferior hold.*

1. $\limsup_{n\to\infty} a_n \geq \liminf_{n\to\infty} a_n$.
2. $\limsup_{n\to\infty} a_n = \liminf_{n\to\infty} a_n = A$ *if and only if* $\lim_{n\to\infty} a_n = A$. *(This holds even if* A *is infinite.)*
3. *If* $\limsup_{n\to\infty} a_n = A$ *and* $\limsup_{n\to\infty} b_n = B$*, then* $\limsup_{n\to\infty}(a_n + b_n) \leq A + B$.
4. *If* $\liminf_{n\to\infty} a_n = A$ *and* $\liminf_{n\to\infty} b_n = B$*, then* $\liminf_{n\to\infty}(a_n + b_n) \geq A + B$.
5. *If* $\limsup_{n\to\infty} a_n = A$ *and* $k > 0$*, then* $\limsup_{n\to\infty}(ka_n) = kA$.
6. *If* $\liminf_{n\to\infty} a_n = A$ *and* $k > 0$*, then* $\liminf_{n\to\infty}(ka_n) = kA$.

There is a nice corollary to the second part of this proposition. Suppose a sequence diverges by oscillation. By the proposition, this will happen exactly when the limit inferior and the limit superior are different. This can only happen if one or more of the following occurs: (i) the sequence is unbounded from above and below, (ii) the sequence is unbounded but not properly divergent, (iii) the sequence has multiple finite subsequential limit points. The third case is the most illustrative—the terms of the sequence are clustered around multiple points, getting close to both (or all) of them but moving back and forth from limit point to limit point. This motivates the term "divergent by oscillation." Viewing increasing without bound as "getting close to ∞," we can interpret all three cases in this vein. An oscillatory sequence has terms that get close to multiple targets, but *oscillate* among the targets.

1.1.2 Infinite Series

Pseudo-Definition 1.33 *A (real) infinite series is obtained by taking the terms of an infinite sequence and connecting them with plus signs rather than commas. The resulting expression is written*

$$\sum_{n=0}^{\infty} a_n$$

or occasionally

$$a_0 + a_1 + a_2 + \cdots,$$

where the index of summation is allowed to begin at values other than 0.

Definition 1.34 *With each infinite series $\sum_{n=0}^{\infty} a_n$ we associate two distinct sequences. First is the sequence of terms, which is simply $\{a_n\}_{n=0}^{\infty}$. Second, the sequence of partial sums $\{s_k\}_{k=0}^{\infty}$ is defined by*

$$\begin{aligned}
s_0 &= a_0 \\
s_1 &= a_0 + a_1 \\
s_2 &= a_0 + a_1 + a_2 \\
&\vdots \\
s_k &= a_0 + a_1 + a_2 + \ldots + a_k = \sum_{n=0}^{k} a_n.
\end{aligned}$$

We define the series $\sum_{n=0}^{\infty} a_n$ to be convergent if and only if $\lim_{k\to\infty} s_k$ exists (and is finite). We define the series $\sum_{n=0}^{\infty} a_n$ to be divergent (not convergent) if and only if $\lim_{k\to\infty} s_k = \infty$ or $\lim_{k\to\infty} s_k = -\infty$ or $\lim_{k\to\infty} s_k$ fails to exist.

Definition 1.35 *In the case that $\sum_{n=0}^{\infty} a_n$ converges to a limit S (i.e., $\lim_{n\to\infty} s_n = S$), we call this limit the sum of the series and write*

$$\sum_{n=0}^{\infty} a_n = S.$$

We may also extend this convention to the divergent cases where

$$\lim_{k\to\infty} s_k = \infty \ (or\ -\infty)$$

and write

$$\sum_{n=0}^{\infty} a_n = \infty \ (or\ -\infty).$$

The reader should notice that the $\sum$ symbols that appear here do not actually represent summation in the ordinary finite sense, but in fact are a special case of the limit concept treated earlier. In particular, we can work with these in all the same ways we worked with ordinary limits. The results of Proposition 1.8 are "inherited" by sums of series, as summarized here.

Proposition 1.36 (Basic Properties of Series) *The following properties of series hold.*

1. *A convergent series has a unique sum.*
2. *A convergent series does not also diverge.*
3. *If $\sum_{n=0}^{\infty} a_n = A$, then $\sum_{n=0}^{\infty}(ka_n) = kA$. This includes the case when A is infinite, provided $k \neq 0$.*
4. *If $\sum_{n=0}^{\infty} a_n = A$ and $\sum_{n=0}^{\infty} b_n = B$, then $\sum_{n=0}^{\infty}(a_n + b_n) = A + B$. This includes the case when A or B is infinite, provided A and B are not both infinite and opposite in sign.*
5. *$\sum_{n=0}^{\infty} a_n$ and $\sum_{n=N}^{\infty} a_n$ have the same limit behavior for any N, except they will in general have different sums in the case when both converge. The latter series is called a tail for the series $\sum_{n=0}^{\infty} a_n$, and any tail contains all the information about the limit behavior of a series. In the case when $\sum_{n=0}^{\infty} a_n$ converges or diverges to ∞, $\sum_{n=0}^{\infty} a_n = \sum_{n=0}^{N-1} a_n + \sum_{n=N}^{\infty} a_n$.*

1.2 Special Series

1.2.1 Geometric Series

Example 1.37 *Analyze the behavior of the infinite series*

$$1 + \frac{1}{2} + \frac{1}{4} + \frac{1}{8} + \frac{1}{16} + \cdots = \sum_{n=0}^{\infty} \left(\frac{1}{2}\right)^n.$$

Solution. By definition, the behavior of this series is just the behavior of its sequence of partial sums. We evaluate by hand the first few of these.

$$\begin{aligned}
s_0 &= 1 = 2 - 1 \\
s_1 &= 1 + 1/2 = 3/2 = 1.5 = 2 - 1/2 \\
s_2 &= 1 + 1/2 + 1/4 = 7/4 = 1.75 = 2 - 1/4 \\
s_3 &= 1 + 1/2 + 1/4 + 1/8 = 15/8 = 1.875 = 2 - 1/8
\end{aligned}$$

We see that, in general, these partial sums appear to be approaching 2. In fact a careful look at these special cases may lead us to suspect the general pattern

$$s_n = 2 - \frac{1}{2^n}.$$

We could try to show this by induction, but instead we will use a clever trick.

$$s_n = 1 + 1/2 + 1/4 + 1/8 + \cdots + 1/2^n$$
$$2s_n = 2 + 1 + 1/2 + 1/4 + \cdots + 1/2^{n-1}$$

Now, all the terms except two are the same in the right-hand sides of these two equations. Subtracting the first from the second,

$$\begin{aligned} s_n &= 2s_n - s_n \\ &= (2 + 1 + 1/2 + 1/4 + \cdots + 1/2^{n-1}) - (1 + 1/2 + 1/4 + 1/8 + \cdots + 1/2^n) \\ &= 2 - 1/2^n, \end{aligned}$$

as desired. Finally,

$$\sum_{n=0}^{\infty} \left(\frac{1}{2}\right)^n = \lim_{n\to\infty} \left(2 - \frac{1}{2^n}\right) = 2.$$

We should pause for a moment here. This example is classical, and many readers will have seen it, even those who may not have seen the definition of an infinite series. We should really be careful to understand what it is we are doing when we claim to have added an infinite collection of numbers. Although in giving our formal definitions, we have spoken of interpreting a sum of an infinite sequence of terms, this is the first concrete example. We are familiar with adding in the finite sense, but how do we *know* what we should mean by this infinite sum? There is no question of "proving" that the definition we gave is right, since right and wrong are not applicable to a definition. (In a sense, of course, we could define a sum of a series any way we pleased.)[1] Although we have seen in Proposition 1.36 that this definition of infinite series convergence has some properties that we would expect, and we will go further in this direction in Chapter 4; on a practical level the authors realize we are asking a lot of the reader's intuition. To that end, we make this example more concrete yet. In Figure 1.1 we begin with a 2 by 1 rectangle and divide it into various rectangles having areas 1, 1/2, 1/4, ..., that is, the terms of the series above. We notice that they fit together to make the larger rectangle, whose area we can compute to be 2. Empirically, then, if the sum of this series should have any meaning at all, our intuition would demand that the sum be 2. So in this case at least our definition gives us an answer that we "like."

The technique that we used to sum this series depended essentially on the fact that multiplying by 2 and subtracting eliminated all the intermediate terms. This, in turn, was successful because each term differed from the following term by the same fixed ratio. Evidently we could use the same technique to sum any series in which the terms have a constant ratio. We capture this tentative observation in a definition and a theorem.

[1] Indeed, there do exist other related notions of summing series with interesting properties in their own right.

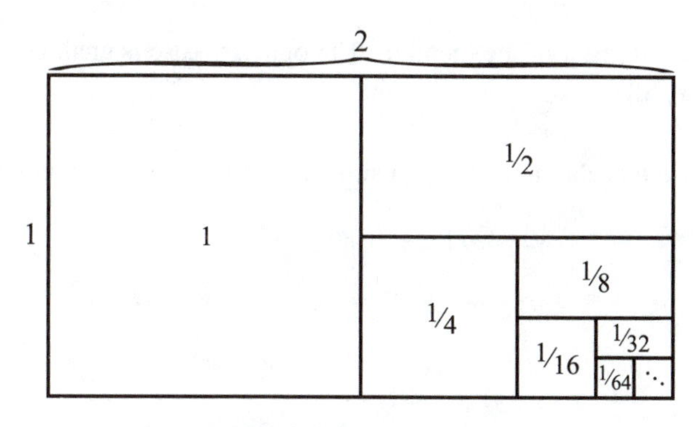

FIGURE 1.1
A Geometric Series of Areas

Definition 1.38 *A geometric series is a series $\sum_{n=0}^{\infty} a_n$ with $a_n = a_0 r^n$ for some number r. r is called the common ratio of the series and is, when $r \neq 0$, the ratio of any term to its immediate predecessor (i.e., a_{n+1}/a_n).*

Theorem 1.39 *Let $\sum_{n=0}^{\infty} ar^n$ be a geometric series with $a \neq 0$.*

1. *If $-1 < r < 1$, then the series converges, and the sum is $a/(1-r)$.*
2. *If $|r| \geq 1$, then the series diverges.*

Proof 1. Note that the theorem is trivial in the cases $r = 0$ and $r = 1$ (in the latter case, all the terms are the same), so suppose $r \neq 0$. As usual let $\{s_k\}$ be the sequence of partial sums for the geometric series. Then we claim

$$s_n = a\frac{1-r^{n+1}}{1-r}.$$

We prove this claim by induction. In the case $n = 0$, we have

$$s_0 = a_0 = a = a(1-r)/(1-r),$$

as desired. Now suppose that, for some $n = k \geq 0$,

$$s_k = a\frac{1-r^{k+1}}{1-r};$$

then for $n = k+1$,

$$s_{k+1} = s_k + a_{k+1} = a\frac{1-r^{k+1}}{1-r} + ar^{k+1} = \frac{a - ar^{k+1} + ar^{k+1} - ar^{k+2}}{1-r} = a\frac{1-r^{k+2}}{1-r},$$

completing the induction. Now, in the case $|r| < 1$, we notice that

$$\lim_{n\to\infty} s_n = \lim_{n\to\infty} a\frac{1-r^{n+1}}{1-r} = \frac{a}{1-r}\lim_{n\to\infty}(1-r^{n+1}) = \frac{a}{1-r},$$

showing the first statement of the theorem. The opposite case is analogous—we note that $\{r^n\}$, and hence s_n, diverges.

Proof 2. We prove this claim in a fashion similar to our earlier development.

$$s_n = a(1 + r + r^2 + \cdots + r^n)$$
$$rs_n = a(r + r^2 + \cdots + r^n + r^{n+1}).$$

Therefore we have that $(1 - r)s_n = a(1 - r^{n+1})$ and if $|r| < 1$, then

$$\lim_{n\to\infty} s_n = \lim_{n\to\infty} a\left(\frac{1 - r^{n+1}}{1 - r}\right) = \frac{a}{1 - r}.$$

If $|r| \geq 1$ clearly $\lim_{n\to\infty}(s_n)$ does not exist.

The above theorem has many implications in the theory of infinite series, for two basic reasons. First, many naturally arising series in applications turn out to be geometric, or combinations of geometric series. Second, the well-understood behavior of geometric series will become the basis of two major tests in Section 1.4.

We also remark that we can use this formula even when the series is not indexed to begin at 0. By a similar argument that the reader can verify, we have

$$\sum_{n=k}^{\infty} ar^n = \frac{ar^k}{1 - r}.$$

This can be remembered as the single pseudo-formula

$$\sum \text{(Geometric Series)} = \frac{\text{first term}}{1 - \text{common ratio}},$$

which holds regardless of how the series is indexed.

Example 1.40 *Analyze the behavior of the infinite series*

$$8 + \frac{32}{3} + \frac{128}{9} + \cdots = \sum_{n=0}^{\infty} 8\left(\frac{4}{3}\right)^n.$$

Solution. We might first find partial sums:

$$s_0 = 8$$
$$s_1 = 8 + 32/3 = 56/3$$
$$s_2 = 8 + 32/3 + 128/9 = 296/9$$

and discover no obvious pattern. In fact, the partial sums appear to be increasing rapidly. Instead we notice that this series is in fact a geometric series with common ratio $r = 4/3$. Since $r > 1$, Theorem 1.39 shows that the series diverges.

Example 1.41 *Interpret the repeating decimal* 0.424242424242... *as an infinite series whose sum is a rational number.*

Solution. We can write the given number, where 42 repeats infinitely often, in the form

$$0.4242424242\ldots = 0.42 + 0.0042 + 0.000042 + \cdots = \sum_{n=0}^{\infty}(0.42)(0.01)^n,$$

so we see that the series is geometric with $a = 0.42$, $r = 0.01$. By Theorem 1.39, the series converges to a sum of

$$\frac{0.42}{1-0.01} = \frac{0.42}{0.99} = \frac{42}{99} = \frac{14}{33}.$$

Therefore, we conclude that the natural meaning to give to the repeating decimal 0.424242424242... must be 14/33. (The reader should notice that every repeating decimal can be interpreted as an initial decimal, then a purely repeating part that can be considered as a geometric series; the reader may wish to verify that this corresponds exactly to the algorithm we all learned in school for converting certain decimals to fractions. As an example, if $x = 0.99999\ldots$ then $10x = 9.9999\ldots$, hence $9x = 10x - x = 9$ thus $x = 1$.)

1.2.2 Telescoping Series

Example 1.42 *Analyze the behavior of the series*

$$\frac{1}{2} + \frac{1}{6} + \frac{1}{12} + \frac{1}{20} + \cdots = \sum_{n=1}^{\infty} \frac{1}{n^2+n}.$$

Solution 1. We might first take partial sums,

$$\begin{aligned}
s_1 &= 1/2 \\
s_2 &= 1/2 + 1/6 = 4/6 = 2/3 \\
s_3 &= 1/2 + 1/6 + 1/12 = 9/12 = 3/4 \\
s_4 &= 1/2 + 1/6 + 1/12 + 1/20 = 16/20 = 4/5,
\end{aligned}$$

and discover the pattern $s_n = n/(n+1)$. To prove this conjecture we can use induction. The base cases have just been taken care of, so we move to the inductive case. Assume that $s_k = k/(k+1)$. Then we can write

$$\begin{aligned}
s_{k+1} = s_k + a_{k+1} = s_k + \frac{1}{(k+1)^2+k+1} &= \frac{k}{k+1} + \frac{1}{k^2+3k+2} \\
= \frac{k(k+2)+1}{k^2+3k+2} = \frac{(k+1)^2}{(k+1)(k+2)} &= \frac{k+1}{k+2},
\end{aligned}$$

completing the induction. We conclude that $s_n = n/(n+1)$. Finally,

$$\sum_{n=1}^{\infty} \frac{1}{n^2+n} = \lim_{n\to\infty} s_n = \lim_{n\to\infty} \frac{n}{n+1} = 1.$$

Solution 2. We first manipulate the terms of the series, to make the pattern in the partial sums more obvious.

$$a_n = \frac{1}{n^2+n} = \frac{1}{n(n+1)} = \frac{1}{n} - \frac{1}{n+1}$$

Now we can quickly see

$$\begin{aligned}
s_1 &= 1 - 1/2 \\
s_2 &= (1-1/2) + (1/2-1/3) = 1 - 1/3 \\
s_3 &= (1-1/2) + (1/2-1/3) + (1/3-1/4) = 1 - 1/4 \\
&\vdots \\
s_n &= (1-1/2) + (1/2-1/3) + \cdots + (1/n - 1/(n+1)) = 1 - 1/(n+1).
\end{aligned}$$

Notice that we can directly compute any partial sum by realizing that most of the terms cancel with one another, collapsing the many-term sum into a simple sum. Mathematicians say that this series *telescopes* or *collapses*, the former term referring to the way an old-fashioned telescope folds up, with each segment disappearing as it fits inside the next. As in Solution 1, we again see $\lim_{n\to\infty} s_n = 1$.

This telescoping behavior is central to many series whose sum we can actually compute numerically. (This includes the geometric series of Section 1.2.1; can you see how to rewrite a geometric series so that it telescopes?) Since using the definition of infinite series directly requires a precise knowledge about the partial sums, we must use some strategy to collapse the nonclosed definition of the partial sum into a closed form. The preceding example is probably the most common telescoping series in the literature, and indeed the telescoping behavior is very commonly accomplished by decomposing the terms of a series into partial fractions. However, the reader should be alert to use this technique, even on series where the decomposition is less immediately obvious.

Example 1.43 *Analyze the behavior of the series*

$$1 + \frac{1}{\sqrt{2}+1} + \frac{1}{\sqrt{3}+\sqrt{2}} + \frac{1}{\sqrt{4}+\sqrt{3}} + \cdots = \sum_{n=0}^{\infty} \frac{1}{\sqrt{n+1}+\sqrt{n}}.$$

Solution. We first rationalize the denominator to make the terms of the series more workable.

$$a_n = \frac{1}{\sqrt{n+1}+\sqrt{n}} = \frac{\sqrt{n+1}-\sqrt{n}}{(\sqrt{n+1}+\sqrt{n})(\sqrt{n+1}-\sqrt{n})}$$
$$= \frac{\sqrt{n+1}-\sqrt{n}}{(n+1)-n} = \sqrt{n+1}-\sqrt{n}$$

Again we can find the telescoping behavior

$$s_0 = \sqrt{1}-\sqrt{0} = 1$$
$$s_1 = (\sqrt{1}-\sqrt{0}) + (\sqrt{2}-\sqrt{1}) = \sqrt{2}$$
$$s_2 = (\sqrt{1}-\sqrt{0}) + (\sqrt{2}-\sqrt{1}) + (\sqrt{3}-\sqrt{2}) = \sqrt{3}$$
$$\vdots$$
$$s_n = (\sqrt{1}-\sqrt{0}) + (\sqrt{2}-\sqrt{1}) + (\sqrt{3}-\sqrt{2}) + \cdots + (\sqrt{n+1}-\sqrt{0}) = \sqrt{n+1},$$

which gives us all the partial sums. Applying the definition of series convergence and divergence,

$$\sum_{n=0}^{\infty} \frac{1}{\sqrt{n+1}+\sqrt{n}} = \lim_{n\to\infty} \sqrt{n+1} = \infty,$$

so that the series diverges to ∞.

1.3 Intuition and Infinity

1.3.1 Zeno's Paradox

The notion of infinity and of infinite processes has been confusing on an intuitive level throughout mathematical history. The reader is almost certainly familiar with Zeno's First Paradox, or at least an unsophisticated interpretation of it, as one "application" of the idea of infinite series. In a sense this is fortunate because the reader knows a bit of the struggle that early intellectuals had with the infinite. Yet, as Pope said, a little knowledge is a dangerous thing, and people have gone on to claim that Zeno's arguments actually show lots of things that are ridiculous. So we should go no further before discussing Zeno's claims in some depth—what they imply and what they *don't* imply. In deference to ancient Greek tradition, we do this in the form of a dialogue between Alice, a modern mathematician, and Zed, descendant and follower of Zeno. Zed's errors represent common and very reasonable misconceptions about the nature of infinite processes. (Zed's second speech is the content of Zeno's First Paradox, in more or less the form that Zeno would have argued it.)

ALICE: You say you think that motion is impossible?

ZED: That's right, just as Zeno explained it. People have said that he was wrong, but I've never understood why. Suppose that a runner, let's say Achilles—

ALICE: Let's say Alice.

ZED: Okay, let's say Alice. Suppose you want to run some distance, let's say a mile. In order to run that mile, you must first cover half the distance. After that, you must go another quarter-mile to get half-way to the finish line. After crossing each half-way point, there is always another half-way point to cross. Since you have an infinite number of half-way points to cross, it would take you forever to run the mile—you couldn't actually do it. So motion is impossible.

ALICE: You can't possibly believe that.

ZED: Why not?

ALICE: For one thing, you've seen me run a mile. And you *know* that motion is possible. I would trust my senses and what I can observe more than an iffy verbal argument any day.

ZED: Well, so would I. I guess that Zeno's work just goes to show that you can't trust reasoning about infinity.

ALICE: Not exactly. You can't trust *bad* reasoning about infinity, any more than you can trust bad reasoning about anything else.

ZED: But Zeno's argument doesn't seem like bad reasoning. It's perfectly good logic.

ALICE: Well, it's sometimes hard to tell whether English speech is good logic or bad logic. Let's try to be as precise as we can about what you and Zeno are saying. To run the mile, we must first accomplish the subtask T_1 of going half the distance, then the subtask T_2 of travelling to the next halfway point, and so on. Our original task, then, breaks up into a sequence of subtasks

$$T_1, T_2, T_3, \dots .$$

Now why don't you think I can ever do all of these tasks? Be as clear as you can be.

ZED: Well, each of the tasks T_i will take you a certain positive amount of time t_i. The whole task would take forever because you would have to add an infinite number of positive numbers.

ALICE: But we can't add an infinite number of numbers like that—

ZED: That's just my point!

ALICE: Not quite. You want to conclude that the total time is actually *infinite*, which suggests that you have some way of interpreting the sum. I say we can't add them all directly because addition doesn't make sense that way; we'd have to use infinite series. If we have a convergent infinite series then good fortune has fallen our way.

ZED: Hmmm. That's true. Okay, then we'll use a series. The total time should be

$$\sum_{i=1}^{\infty} t_i .$$

That must be infinite because, um, because, it just has to be.

ALICE: You're making the same assumption that Zeno did; indeed, that Greek mathematicians all did. You're assuming that any series of positive terms has to diverge. They never state that assumption because they did not have the vocabulary at that point in the history of mathematics. But basically that's the fallacious "fact" that you and Zeno are invoking.

ZED: Seems reasonable.

ALICE: But we know that isn't true. Back in Example 1.42, we found a convergent series.

ZED: Oh. Right. Okay, so maybe Zeno's reasoning about infinity wasn't valid. But why should I believe that your definition of infinite series is any more sensible? It sounds fine, but Zeno's Paradox sounded fine to me yesterday.

ALICE: That's a good question. Of course, as a mathematical definition, it's just as good as any other definition. Strictly speaking, it doesn't have to make "sense" as long as it isn't evidently inconsistent. But in this case, the notion of summing an infinite series does have a meaningful interpretation, even in Zeno's example. Let's suppose that I can run about 10 mph, and I want to know how long it would take to run the mile.

ZED: Six minutes!

ALICE: Don't interrupt. You've followed Zeno this far, let's follow him all the way, and this time use the definition to try to sum the series. The first task T_1 involves running $1/2$ mile, so I can do it in $(1/2 \text{ mile})/(10 \text{ mph}) = 1/20 \text{ hours} = 3$ minutes. Then $t_1 = 3$ minutes. T_2 is to run half as far, so $t_2 = 1.5$ minutes. In general, T_n is to run $1/2^n$ miles, and $t_n = 6/2^n$ minutes. We want the summation

$$\sum_{n=1}^{\infty} t_n = \sum_{n=1}^{\infty} \frac{6}{2^n},$$

which is geometric with first term 3 and common ratio $1/2$. Using what we know about geometric series, this sum is $3/(1-1/2)$.

ZED: Six minutes, as I said in the first place.

ALICE: Right. Reasoning about infinite series will give you the answers you were expecting to questions you could answer already; however, in some cases, these definitions will let you solve problems you couldn't solve without resorting to an infinite process.

Note: Let us return to Alice's next to last statement and observe that the initial half mile required three minutes running time and likewise three minutes running time for the remaining half mile. If we view the one mile run as an infinite number of tasks, then the required time would be $3 + \frac{3}{2} + \frac{3}{4} + \frac{3}{8} + \cdots$ minutes. These two observations are often represented by the equation $3 + 3 = 3 + \frac{3}{2} + \frac{3}{4} + \frac{3}{8} + \cdots$. Zeno saw this as paradoxical. Zeno assumed a priori that no actual infinity exists, hence no infinite process could be completed and consequently no equivalence between a finite and infinite process could be allowed.

1.3.2 Series vs. Addition

As the previous section indicates, it is possible to get in a lot of logical trouble if we are not careful about how we manipulate and interpret infinite series. We are both helped and hindered by the connection between infinite series and addition. For most readers, addition was the first mathematical operation learned. Ever since kindergarden ("If Joelle has five apples and...") we have dealt with addition tables and a widening variety of addition problems and applications. In fact, because of this depth of personal experience, the properties of addition are very seldom thought about in any sort of explicit way. We just "know" how addition works. When working with series, we have to remember that we are "overloading" the $\sum$ and $+$ operators by defining series notation as we have. Despite being defined by analogy with and in terms of ordinary addition, series summation is *not* ordinary addition. In the equations

$$5+2=7 \text{ and } 1/2+1/6+1/12+1/20+\cdots=1,$$

the $+$ represents a different operation with different properties.

We know that series summation is not ordinary addition because, ultimately, addition is a binary operation. We only really know how to add numbers two at a time. Poppycock, you may be saying, because we all have seen sums like

$$1+2+3+4+5+6+7+8+9+10=55.$$

However, this "10-ary" form of addition is really constructed by inductively using binary addition. When interpreting a summation like this, we really say, "$1+2=3, 3+3=6, 6+4=10, \ldots$," adding two at a time. We cannot use this recursive process to define an ∞-ary addition. However, notice that the definition of summation does parallel this process—the partial sums represent all the intermediate answers we *would* get by blithely trying to add the infinite list of terms.

It is true that many of the familiar properties of addition, now second nature due to the reader's extensive personal experience with adding, apply in an analogous form to infinite series. For example, adding two convergent series, term by term, will form a new convergent series, and its sum will be the sum of the sums of the original series. However, nothing should go without saying. The reader should try now to prove the linearity property for series directly from the definition as a straightforward exercise. There are two very common errors when working with series that stem from abusing properties of addition, and they are worth mentioning explicitly. The following two properties should not be applied fast and loose to infinite series.

Addition is commutative. We know that we can exchange the positions of terms in a finite summation to our heart's content. However, at least *a priori*, order does matter in the definition of the partial sums. If we take a given series $\sum_{n=0}^{\infty} a_n$ and make a new series

$$a_{100}+a_1+a_{435}+a_{21}+a_6+\cdots,$$

obtained by using each of the a_i in some order, the resulting series is called a *rearrangement* of the original series. In general, these two series do not necessarily both converge or both diverge, and even if both converge they may have different sums. In Chapter 4, we will learn more about the relationship between a series and its rearrangement.

Addition is associative. In a finite summation, we may insert and rearrange parentheses as desired. However, in moving from the expression

$$a_0 + a_1 + a_2 + a_3 + a_4 + a_5 + \cdots$$

to the expression

$$(a_0 + a_1) + (a_2 + a_3) + (a_4 + a_5) + \cdots,$$

we have changed it substantially. The $+$ symbols inside the parentheses now indicate ordinary addition, whereas they indicated series summation in the first expression. A series obtained by collapsing some adjacent terms into single terms (that is, inserting parentheses) is called a *grouping* of the original series. Again, we will see in Chapter 4 what sort of relationship a series has to its groupings. A particularly famous abuse of grouping and regrouping was used to show that $0 = 1$. The argument went as follows:

$$\begin{aligned}1 &= 1 + 0 + 0 + 0 + 0 + \cdots = 1 + (-1 + 1) + (-1 + 1) + (-1 + 1) + \cdots \\ &= 1 - 1 + 1 - 1 + 1 - 1 + 1 - 1 + \cdots = (1 - 1) + (1 - 1) + (1 - 1) + \cdots \\ &= 0 + 0 + 0 + 0 + \cdots = 0\end{aligned}$$

The creator of this reasoning error claimed it showed the existence of God. The authors of this book claim rather that it shows the importance of remembering the difference between finite addition and "infinite addition" by way of infinite series.

1.3.3 Qualitative Behavior of Series

Before undergoing any heavy analysis of infinite series, it may be worth taking a moment to imagine, at least in broad terms, what sorts of behavior infinite series can have. In thinking about series for the purpose of this section, we can for the most part disregard actual numbers and think only about behavior. Visualize a series as a sequence of steps on a number line, beginning at 0 before any terms are added. As each successive term a_n is added, the point marking the running total moves to the right or the left according as a_n is positive or negative.

First consider the case in which all the terms are positive. Then the running total must step from left to right, never moving back to the left. Then there are really only two things that can happen to the running total. Either the partial sums will increase without bound to the right, meaning that the series will diverge to ∞, or it will move to the right slowly enough that the partial sums will cluster at a point L, which would mean that the series converges to L.[2]

[2]The reader with a particularly vivid imagination may imagine that the running total moves to the right, staying always less than some bound, but without actually reaching any limit. However, this can never happen—it is a consequence of a property of the real numbers called "completeness." In this treatment, the property of completeness is assumed when we state that every set has a supremum and an infimum. We refer the reader to any text on analysis for more details.

In the case that the series may have terms of both signs, then the running total may move in either direction. Again, the point may move off steadily to the right without limit (in which case the series sums to ∞), or it may move off steadily to the left without limit (in which case the series sums to $-\infty$). Also as before, the running total may get closer and closer to some specific point, in which case the series converges. However, we can also imagine some other cases: the running total may move to the right to a certain number A, then back to the left to approach B, then back toward A, and so on. We can think of this hypothetical series as going back and forth between A and B. Although there are many variations on this behavior, it seems that any behavior other than convergence or divergence to $\pm\infty$ must involve some sort of back and forth motion of the partial sum. It is this qualitative description that motivates the term "oscillation" defined earlier.

In every case, then, when the series converges, the partial sums must eventually all be arbitrarily close indeed to the limit. If the partial sums eventually all stay in small intervals centered on the limit, then no step (read: term) can be large enough to move us outside the interval. The following theorem makes this explicit and rigorous.

Theorem 1.44 *Let $\sum_{n=0}^{\infty} a_n$ be a convergent series. Then*

$$\lim_{n\to\infty} a_n = 0.$$

Proof. Define as usual the sequence of partial sums

$$s_k = \sum_{n=0}^{k} a_n.$$

By hypothesis, the series converges to a sum, which we call S. By definition we have

$$\lim_{n\to\infty} s_n = S.$$

Now, we notice that $s_n = a_0 + a_1 + \cdots + a_{n-1} + a_n = (a_0 + a_1 + \cdots + a_{n-1}) + a_n = s_{n-1} + a_n$. Thus

$$\lim_{n\to\infty} a_n = \lim_{n\to\infty} (s_n - s_{n-1}) = \lim_{n\to\infty} s_n - \lim_{n\to\infty} s_{n-1} = S - S = 0.$$

The simplicity of the statement of this theorem and the brevity of its proof may lead the reader to underestimate its value. To put a sharper point on the result, we rephrase it in contrapositive form.

Corollary 1.44.1 (Divergence Test) *Suppose that $\sum_{n=0}^{\infty} a_n$ is a series and*

$$\lim_{n\to\infty} a_n \neq 0.$$

(That is, the limit is not zero, which includes the possibility that the limit does not exist.) Then $\sum_{n=0}^{\infty} a_n$ diverges.

This theorem forms an efficient test to show that certain series diverge. It is generally straightforward to apply, and if the criterion $\lim_{n\to\infty} a_n = 0$ fails, computing the partial sums and any other analysis become superfluous.

On the other hand, we may wonder now about the converse of this theorem, whether any series whose terms tend to 0 must in fact converge. On the face of it, it seems that some variation on the proof of Theorem 1.44 might verify the converse. Theorem 1.39 shows that, at least for geometric series, the condition that the terms tend to 0 (i.e., $|r| < 1$) is necessary and sufficient for the series to converge. Unfortunately this is not true in general. In case you missed it the first time, *this is not true in general.* It is a very common error among students to apply the nonexistent converse of Theorem 1.44 to "show" convergence. Example 1.43, for example, offers a series that diverges even though its terms tend to 0. Loosely speaking, this series diverges because, although its terms go to 0, they decrease very slowly; a series of positive terms will converge if its terms tend to 0 fast enough. Many of the interesting theorems in the theory of infinite series deal with the question of how to determine if a series whose terms go to 0 converges or diverges.

However, there is another way to think about the terms of a sequence "eventually getting very close together." This concept, named for Cauchy, is very useful because it does provide a necessary and sufficient condition.

Definition 1.45 *A sequence $\{a_n\}$ is called a Cauchy sequence if for every $\epsilon > 0$ there exists an integer N such that for all $n_0, n_1 \geq N$, we have*

$$|a_{n_1} - a_{n_0}| < \epsilon.$$

Theorem 1.46 (Cauchy Criterion) *A sequence converges if and only if it is a Cauchy sequence.*

Proof. To show the "if" case, assume that for every $\epsilon > 0$, there exists an integer N such that for all $n_0, n_1 \geq N$, we have

$$|a_{n_1} - a_{n_0}| < \epsilon.$$

Write $L = \liminf_{n\to\infty} a_n$, $R = \limsup_{n\to\infty} a_n$. If $L = R$, then $\{a_n\}$ converges to their common value. Assume for the sake of contradiction that $R - L = E > 0$. By hypothesis, we can find an N such that for every $n_1, n_0 \geq N$, $\left|a_{n_1} - a_{n_0}\right| < E/3$. Since L and R are limit points, we can find infinitely many indices i, j such that $|L - a_i| < E/3$, $|R - a_j| < E/3$. In particular we can choose $i, j > N + 1$, so that $|a_i - a_j| < E/3$. Then

$$R - L = |R - L| \leq |R - a_j| + |a_j - a_i| + |L - a_i| < E/3 + E/3 + E/3 = E = R - L,$$

a contradiction.

To show the "only if" case, suppose that $\{a_k\}$ converges to a limit a. Let $\epsilon > 0$ be given. Then there exists a positive integer N such that, for all $n > N$, $|a_n - a| < \epsilon/2$. Then, for all $n_1 \geq n_0 \geq N + 1$, we have

$$\left|a_{n_1} - a_{n_0}\right| \leq |a_{n_1} - a| + |a_{n_0} - a| < \epsilon.$$

1.4 Basic Convergence Tests

As the reader may have noticed, all the series that we have been able to sum exactly have been "special." In general, given a series, even one with very simply defined terms, such as $\sum_{n=1}^{\infty} 1/(n^2+1)$, we cannot write the partial sums in an elementary closed form. As a result, it is usually too ambitious to expect to evaluate a series directly by the definition. This is typical in mathematics. For example, the definition of integral makes its connection to area and other applications much more evident than would, say, a definition by way of the Fundamental Theorem of Calculus, but is completely impractical to apply; how many definite integrals has the reader completed by decomposing an area into millions of tiny rectangles? In the same way we now develop theorems that allow us to reason about the behavior of series without using the well-motivated but hopelessly impractical definition.

In this section, we will concern ourselves exclusively with *positive series*, that is, series $\sum_{n=0}^{\infty} a_n$ in which $a_n \geq 0$ for all n. If we wish to require that in fact the terms be positive and not merely nonnegative, we will call such series *strictly positive*. We begin in this way because of the straightforward behavior anticipated in Section 1.3.3; first, we show that such a simple behavior actually occurs.

Proposition 1.47 *Let $\sum_{n=0}^{\infty} a_n$ be a positive series. Then one of the following two cases occurs:*

1. $\sum_{n=0}^{\infty} a_n = A$ *for some* $A \geq 0$*, that is,* $\sum_{n=0}^{\infty} a_n$ *converges.*
2. $\sum_{n=0}^{\infty} a_n = \infty$*, that is,* $\sum_{n=0}^{\infty} a_n$ *diverges to* ∞*.*

Proof. Define $s_k = \sum_{n=0}^{k} a_n$ to be the sequence of partial sums, and observe that

$$s_0 \leq s_1 \leq s_2 \leq \cdots \leq s_k \leq s_{k+1} \leq \cdots.$$

That is, $\{s_k\}$ is an increasing sequence. Hence we are really reinterpreting Fact 1.30 for series; bounded increasing sequences must converge. If $\{s_k\}$ is bounded above, then it converges to its least upper bound A, and $\sum_{n=0}^{\infty} a_n = A$. On the other hand, if $\{s_k\}$ is not bounded above, then $\sum_{n=0}^{\infty} a_n = \infty$.

We have remarked elsewhere that in general one must be careful about the order of the terms in an infinite series. Two series with the same terms in a different order are, in general, two different series with no a priori connection. However, the simplistic behavior of positive series guarantees that in this limited context we can be less careful.

Theorem 1.48 *Let $\sum_{n=0}^{\infty} a_n$ be a positive series and let $\sum_{n=0}^{\infty} b_n$ be a rearrangement of the first series. Then the two series either both converge to the same sum or both diverge properly.*

Proof. Using the previous theorem, we know these series cannot oscillate, so we boldly write

$$\sum_{n=0}^{\infty} a_n = A; \sum_{n=0}^{\infty} b_n = B,$$

allowing that A or B may be ∞. Let s_k be some partial sum of the first series. Since each term $a_0, a_1, \ldots, a_k$ that is represented in the partial sum occurs as some b_{j_k}, we can (by considering $j = \max j_k$) find a partial sum of the second series that contains all the terms in s_k and in general some other terms. Since all the terms are positive,

$$\sum_{n=0}^{k} a_n \leq \sum_{n=0}^{j} b_n \leq B.$$

That is, all the partial sums of the first series are bounded above by partial sums of the second series, hence by the sum of the second series. Since this inequality holds for all partial sums, it continues to hold when we pass to the limit, and $A \leq B$. (This is just our Fact 1.28 in action.) But we can use a symmetric argument to guarantee $B \leq A$. Hence $A = B$ as desired.

(The reader may wish to take a moment to convince himself or herself that a similar argument goes through even when A and B are allowed to be infinite; the important modification is as follows: if $\sum_{n=0}^{\infty} a_n = \infty$ then the partial sums are not bounded, so $\sum_{n=0}^{k} a_n \leq B$ for all k means that B must be infinite.)

1.4.1 Comparison Test

Since we suggested earlier that a positive series will converge if its terms get small "quickly enough," it seems plausible that, if one series has smaller terms than another, the series whose terms are smaller is more likely to converge in some sense. In fact this can be formalized as the most basic convergence test. Notice that this is our first result that can show convergence without computing the partial sums—the Divergence Test can only be used to show divergence.

Theorem 1.49 (Comparison Test) *Let $\sum_{n=0}^{\infty} a_n$ and $\sum_{n=0}^{\infty} b_n$ be positive series, and suppose that $a_n \leq b_n$ holds for all n greater than some integer N.*

1. *If $\sum_{n=0}^{\infty} b_n$ converges, $\sum_{n=0}^{\infty} a_n$ converges also.*
2. *If $\sum_{n=0}^{\infty} a_n$ diverges, $\sum_{n=0}^{\infty} b_n$ diverges also.*

Proof. Because the behavior of any series is determined only by its tail, we may consider without loss of generality that $a_n \leq b_n$ for *all* n. Then, the partial sums for the two series satisfy the relationship

$$\sum_{n=0}^{k} a_n \leq \sum_{n=0}^{k} b_n$$

for all k. Since the right-hand side is bounded above, the left-hand side cannot go to infinity. By Proposition 1.47, $\sum_{n=0}^{k} a_n$ must also converge.

The second statement of the theorem is trivial, because it is the contrapositive of the first part.

Notice that we were able to prove a result about the series using a condition that only held for n sufficiently large. When a property does not necessarily hold for all n, but only for all $n > N$, with N some integer, we say that the property is true *eventually*. As suggested by the above proof, it is often sufficient in the study of infinite series to assume that properties of interest are true only eventually.

In the case where $a_n \le b_n$ for all n, this argument actually shows that this inequality is inherited by the partial sums of the two series, and by extension their sums. In symbols,

$$a_n \le b_n \quad \text{for all} \quad n \Rightarrow \sum_{n=0}^{\infty} a_n \le \sum_{n=0}^{\infty} b_n.$$

In fact this includes the case when either or both series diverge to ∞; this may be considered an abuse of notation, but it is a convenient and insightful one. The reader should also note well that this conclusion does not extend to series with both positive and negative terms, because then the series may have no sums at all, and the summation symbols no numerical interpretations. However, in the case when both series are known *a priori* to converge or diverge properly, this comparison result on their sums holds.

Example 1.50 *Analyze the behavior of the series*

$$1 + \frac{2}{3} + \frac{2}{5} + \frac{2}{9} + \frac{2}{17} + \cdots = \sum_{n=0}^{\infty} \frac{2}{2^n + 1}.$$

Solution. We consider the two series

$$\sum_{n=0}^{\infty} \frac{2}{2^n + 1} < \sum_{n=0}^{\infty} \frac{2}{2^n},$$

where the $<$ sign holds because we have decreased the denominator. However, by Theorem 1.39, the right-hand series is a convergent geometric series. Hence the original series converges by the Comparison Test.

Example 1.51 *Analyze the behavior of the series*

$$1 + \frac{1}{\sqrt{2}} + \frac{1}{\sqrt{3}} + \frac{1}{\sqrt{4}} + \frac{1}{\sqrt{5}} + \cdots = \sum_{n=1}^{\infty} \frac{1}{\sqrt{n}}.$$

Solution. We first consider the two series

$$\sum_{n=1}^{\infty} \frac{1}{2\sqrt{n}} \ge \sum_{n=1}^{\infty} \frac{1}{\sqrt{n} + \sqrt{n+1}},$$

where the $\ge$ sign holds because we have increased the denominator this time. But we already know that the right-hand series diverges—we have seen this series before in

Example 1.43. By the Comparison Test, the left-hand side diverges also. However, the series on the left is just a scalar multiple of the original series, and so they converge and diverge together. The original series diverges.

1.4.2 Limit Comparison Test

A drawback to the preceding theorem is that, if we have two series we wish to compare, we must prove that the inequality goes in the right direction for the terms. In the last example to that section, that was handled by inserting a factor of 2. Such a strategy often works, but finding the correct factor and showing that it actually works can be tedious and not insightful for complicated expressions. We want to be able to say that two series with "similar" terms should have the same behavior.

Theorem 1.52 (Limit Comparison Test) *Let $\sum_{n=0}^{\infty} a_n$ and $\sum_{n=0}^{\infty} b_n$ be two series with $a_n \geq 0, b_n > 0$, and assume that $\lim_{n\to\infty}(a_n/b_n) = L$, allowing the case $L = \infty$.*

1. *If $0 < L < \infty$, then $\sum_{n=0}^{\infty} a_n$ and $\sum_{n=0}^{\infty} b_n$ either both converge or both diverge.*
2. *If $L = \infty$ and $\sum_{n=0}^{\infty} a_n$ converges, then $\sum_{n=0}^{\infty} b_n$ converges also.*
3. *If $L = 0$ and $\sum_{n=0}^{\infty} a_n$ diverges, then $\sum_{n=0}^{\infty} b_n$ diverges also.*

Proof. Let $\sum_{n=0}^{\infty} a_n$ and $\sum_{n=0}^{\infty} b_n$ be two series with $a_n \geq 0$ and $b_n > 0$.

Case 1. $0 < L < \infty$. Let $\epsilon = L/2 > 0$ and choose N sufficiently large that for all $n > N$ we have

$$\left|\frac{a_n}{b_n} - L\right| < \epsilon = \frac{L}{2},$$

or equivalently

$$-\frac{L}{2} < \frac{a_n}{b_n} - L < \frac{L}{2},$$

hence $(\frac{1}{2}L)b_n < a_n < (\frac{3}{2}L)b_n$. Appealing to the (regular) Comparison Test, Theorem 1.49, we see that the first assertion follows.

Case 2. $L = \infty$. In this case we must eventually have $a_n/b_n > 1$, hence $a_n > b_n$, and again Theorem 1.49 ensures that if $\sum_{n=0}^{\infty} a_n$ converges then the "smaller series" $\sum_{n=0}^{\infty} b_n$ converges as well.

Case 3. $L = 0$. In this case we must eventually have $a_n/b_n < 1$, hence $a_n < b_n$, and again Theorem 1.49 guarantees that if $\sum_{n=0}^{\infty} a_n$ diverges then the "larger series" $\sum_{n=0}^{\infty} b_n$ diverges as well.

In deciding whether to use the comparison test or the limit comparison test people generally make the choice depending on their agility in handling inequalities versus handling limits.

Readers comfortable with limits superior and inferior will have no trouble adapting the proof to the following version of the Limit Comparison Test. This is the first of several such "strengthened" tests; in every case the proof of the strengthened version is a natural generalization of the proof of the basic theorem, so we will not give the proofs here. The reader who wishes to skip the strengthened versions may feel free to do so. The reason these tests are stronger is that they can cope with series whose terms "jump around" in which case the desired limits do not exist.

Theorem 1.53 (Limit Comparison Test Strengthened) *Let $\sum_{n=0}^{\infty} a_n$ and $\sum_{n=0}^{\infty} b_n$ be strictly positive series, and suppose that*

$$\liminf_{n\to\infty}(a_n/b_n) = L_1 \quad \text{and} \quad \limsup_{n\to\infty}(a_n/b_n) = L_2.$$

1. *If $L_2 < \infty$ and $\sum_{n=0}^{\infty} b_n$ converges, then $\sum_{n=0}^{\infty} a_n$ converges also.*
2. *If $L_1 > 0$ and $\sum_{n=0}^{\infty} b_n$ diverges, then $\sum_{n=0}^{\infty} a_n$ diverges also.*

Example 1.54 *Analyze the behavior of the series*

$$\sum_{n=1}^{\infty} \frac{2^n + n - 2}{3^n + 4n - 5}.$$

Solution. Let the terms of the original series be a_n, and define a new series

$$\sum_{n=1}^{\infty} b_n = \sum_{n=1}^{\infty} \frac{2^n}{3^n}.$$

Now

$$L = \lim_{n\to\infty} \frac{a_n}{b_n} = \lim_{n\to\infty} \frac{\frac{2^n+n-2}{3^n+4n-5}}{\frac{2^n}{3^n}} = \lim_{n\to\infty} \frac{1 + n2^{-n} - 2(2^{-n})}{1 + 4n3^{-n} - 5(3^{-n})} = 1.$$

By Theorem 1.39, $\sum_{n=1}^{\infty} b_n$ is a convergent geometric series. By the Limit Comparison Test, then the original series converges.

1.4.3 Ratio Comparison Test

The following test is actually no stronger than the ordinary Comparison Test, but it is often more convenient to use in the case when the ratio between consecutive terms is simpler than the terms themselves. This is especially true when the terms are *defined* by products, as in Example 1.56. This test should not be confused with the Ratio Test, a very different technique that we present later.

Theorem 1.55 (Ratio Comparison Test) *Let $\sum_{n=0}^{\infty} a_n$ and $\sum_{n=0}^{\infty} b_n$ be strictly positive series, and suppose that $a_{n+1}/a_n \le b_{n+1}/b_n$ holds eventually.*

1. *If* $\sum_{n=0}^{\infty} b_n$ *converges,* $\sum_{n=0}^{\infty} a_n$ *converges also.*
2. *If* $\sum_{n=0}^{\infty} a_n$ *diverges,* $\sum_{n=0}^{\infty} b_n$ *diverges also.*

Proof. Since the behavior of a series is dependent only on the behavior of its tail, we can assume without loss of generality that $a_{n+1}/a_n \leq b_{n+1}/b_n$ universally. By multiplying these inequalities, we find that

$$\frac{a_n}{a_0} \leq \frac{b_n}{b_0} \Rightarrow a_n \leq \frac{a_0}{b_0} b_n.$$

Now, consider the series with terms $c_n = (a_0/b_0)b_n$. Clearly $\sum_{n=0}^{\infty} c_n$ and $\sum_{n=0}^{\infty} b_n$ have the same behavior. The above inequality guarantees that $a_n \leq c_n$. This theorem, then, is equivalent to the ordinary Comparison Test.

Example 1.56 *Analyze the behavior of the series*

$$\sum_{n=1}^{\infty} a_n,$$

where the terms are given by

$$a_1 = 1, a_n = \frac{1}{4}\frac{2}{5}\frac{3}{6} \cdots \frac{n-1}{n+2}$$

for $n > 1$.

Proof. In addition to the given series, consider also the series

$$\sum_{n=1}^{\infty} b_n = \sum_{n=1}^{\infty} \frac{1}{n(n+1)}.$$

This series is known to converge, as we learned in Example 1.42. The Ratio Comparison Test is a natural thing to consider here because the terms a_n are essentially defined by their ratio. We see that

$$\frac{a_{n+1}}{a_n} = \frac{\frac{1}{4}\frac{2}{5}\frac{3}{6} \cdots \frac{n-1}{n+2}\frac{n}{n+3}}{\frac{1}{4}\frac{2}{5}\frac{3}{6} \cdots \frac{n-1}{n+2}} = \frac{n}{n+3}$$

and

$$\frac{b_{n+1}}{b_n} = \frac{\frac{1}{(n+1)(n+2)}}{\frac{1}{n(n+1)}} = \frac{n}{n+2}.$$

Since

$$\frac{n}{n+3} \leq \frac{n}{n+2},$$

we can apply the Ratio Comparison Test to see that the original series converges.

1.4.4 Integral Test

The description of an infinite series as describing the limit behavior of partial sums is motivated by the following line of thinking: "We can't actually sum an infinite list of terms, but we can add as many as we want, and we can consider the limit to generalize to the infinite case." Compare this with the transition from a proper integral to an improper integral: "We can't actually integrate over an infinite region, but we can integrate over very large regions, and we consider the limit to generalize to the infinite case." This suggests that there may be a connection between the two notions. Pictorially, we can see the relationship between the partial sums (interpreted as area) and the area measured by the integral.

Theorem 1.57 (Integral Test) *Consider a series $\sum_{n=1}^{\infty} a_n$, where $a_n = a(n)$ for some function $a(x)$ defined on $[1, \infty)$ that is continuous, positive, and decreasing. Then the infinite series $\sum_{n=1}^{\infty} a_n$ converges if and only if the improper integral*

$$\int_1^{\infty} a(x)dx$$

converges.

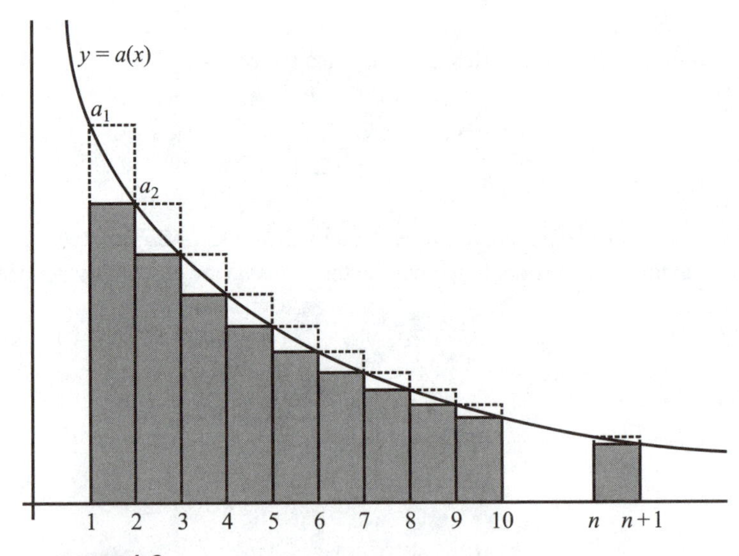

FIGURE 1.2
Partial Sums of an Infinite Series as Upper and Lower Riemann Sums

In the case of convergence, write S for the sum $\sum_{n=1}^{\infty} a_n$ and write I for the integral

$$\int_1^{\infty} a(x)dx.$$

Then $S - a_1 < I < S$.

Proof. Let $\{a_n\}$ and the function a be as described in the hypotheses. Then define two auxiliary functions $\hat{a}$ and $\tilde{a}$ as

$$\hat{a}(x) = a(\lceil x \rceil), \tilde{a}(x) = a(\lfloor x \rfloor).$$

Remark. ($\lceil x \rceil$ and $\lfloor x \rfloor$ denote the round up, round down functions respectively; also known as ceiling and floor functions.) Because $\lceil x \rceil \geq x \geq \lfloor x \rfloor$, the decreasing nature of a guarantees

$$\hat{a}(x) = a(\lceil x \rceil) \leq a(x) \leq a(\lfloor x \rfloor) = \tilde{a}(x).$$

From this we have for each integer k,

$$\int_1^k \hat{a}(x)dx \leq \int_1^k a(x)dx \leq \int_1^k \tilde{a}(x)dx.$$

However, recall that for positive functions an integral measures the area under a curve. Since $\hat{a}, \tilde{a}$ are piecewise constant, this area is made of rectangles of width 1. If we write as usual s_k for the kth partial sum of the series, a little consideration shows that the previous compound inequality can be rewritten

$$s_k - s_1 \leq \int_1^k a(x)dx \leq s_{k-1}.$$

Now by the definition of infinite series (resp. improper integral) the series (resp. integral) will diverge if and only if the $\{s_k\}$ (resp. $\int_1^k a(x)dx$) increase to infinity. If the integral diverges, the second inequality shows that the partial sums also increase to ∞, and the series diverges. Conversely, if the series diverges, the first inequality forces the improper integral to diverge. We conclude that the series and integral must in fact have the same behavior. In the case when the series converges, the claimed inequality $S - a_1 < I < S$ follows from the above inequality by passing to the limit.

With this test we have a decided advantage over the previous tests. The other tests, powerful as they are, require the problem-solver to pull from his or her knowledge bank some other series that stands in some relationship to the given series—that is, the tests relate to things *extrinsic* to the given series. Without access to divine inspiration or voodoo magic, such constructions are sometimes difficult to devise. This test is *intrinsic* to the series. Now, we may argue that in fact we did have to make an inspired choice in the Integral Test, namely to choose the function to integrate. Generally, however, there is only one obvious choice for the function. The tests in the next two sections (1.4.5 and 1.4.6) will be more purely intrinsic.

Example 1.58 (Example 1.43 Revisited) *Analyze the behavior of the series*

$$1 + \frac{1}{\sqrt{2}} + \frac{1}{\sqrt{3}} + \frac{1}{\sqrt{4}} + \frac{1}{\sqrt{5}} + \cdots = \sum_{n=1}^{\infty} \frac{1}{\sqrt{n}}.$$

Solution. In this case $a_n = a(n)$ for the function $a(x) = x^{-1/2}$. The reader can quickly verify that the function a is positive, continuous and decreasing, so the Integral Test guarantees that the infinite series

$$\sum_{n=1}^{\infty} \frac{1}{\sqrt{n}}$$

will behave in the same way as

$$\int_1^{\infty} \frac{1}{\sqrt{x}}\, dx.$$

We evaluate this improper integral in the usual way:

$$\int_1^{\infty} \frac{1}{\sqrt{x}}\, dx = \lim_{B\to\infty} 2\sqrt{x}\big|_1^B = \lim_{B\to\infty} (2\sqrt{B} - 2) = \infty.$$

Since the integral diverges, the series is also divergent to ∞.

Theorem 1.59 (p-series) *Consider the series*

$$\sum_{n=1}^{\infty} \frac{1}{n^p},$$

called a p-series, where p is a real constant.[3]

1. *If $p > 1$, the series converges.*
2. *If $p \leq 1$, the series diverges.*

Proof. In the case where $p \leq 0$, we trivially have divergence by the Divergence Test, so we can assume $p > 0$ in what follows. Under this hypothesis we notice that $1/x^p$ is a monotonically decreasing continuous positive function, so the hypotheses of the Integral Test are satisfied.

Case 1. $p > 1$.

$$\int_1^{\infty} x^{-p}\, dx = \lim_{B\to\infty} \frac{x^{1-p}}{1-p}\bigg|_1^B = \lim_{B\to\infty} \left(\frac{B^{1-p}}{1-p} + \frac{1}{p-1}\right) = 1/(p-1) < \infty,$$

so that the p-series converges.

[3]Saying that p is a fixed constant is strictly speaking redundant, but we wish to emphasize that this result does not extend to the case where $p = p_n$ depends on n, even if $p_n > 1$ for all n.

Case 2. $p = 1$.

$$\int_1^\infty x^{-1}\,dx = \lim_{B\to\infty} \log x|_1^B = \lim_{B\to\infty} \log B = \infty,$$

so that the p-series diverges.

Case 3. $p < 1$.

$$\int_1^\infty x^{-p}\,dx = \lim_{B\to\infty} \left.\frac{x^{1-p}}{1-p}\right|_1^B = \lim_{B\to\infty} \left(\frac{B^{1-p}}{1-p} + \frac{1}{p-1}\right) = \infty,$$

so that the p-series diverges.

The preceding theorem really comes into its own as a powerful tool when combined with the Limit Comparison Test. In fact, we can now analyze the behavior of any series whose terms are given by an algebraic function. The procedure is to compute the overall "degree"[4] of the terms (as a function of n), and to choose a p-series with the same overall degree; two examples should suffice to illustrate the method.

Example 1.60 *Analyze the behavior of the series*

$$\sum_{n=1}^\infty \frac{n^2+n-1}{23n^5 - n^4 + 3n^3 - 9n^2 + 6n - 11}.$$

Solution. The numerator has degree 2 and the denominator has degree 5. Therefore we want a series with degree -3 and choose

$$\sum_{n=1}^\infty b_n = \sum_{n=1}^\infty \frac{1}{n^3},$$

which is a convergent p-series. Using the Limit Comparison Test, we find

$$\begin{aligned}
&\lim_{n\to\infty} \left(\frac{n^2+n-1}{23n^5 - n^4 + 3n^3 - 9n^2 + 6n - 11}\right) \Big/ \left(\frac{1}{n^3}\right) \\
&\quad = \lim_{n\to\infty} \frac{n^5+n^4-n^3}{23n^5 - n^4 + 3n^3 - 9n^2 + 6n - 11} \\
&\quad = \lim_{n\to\infty} \frac{1+n^{-1}-n^{-2}}{23 - n^{-1} + 3n^{-2} - 9n^{-3} + 6n^{-4} - 11n^{-5}} = 1/23,
\end{aligned}$$

and we see that the given series also converges.

---

[4]We are using a fairly liberal interpretation of the notion of degree. For example, it is perhaps not so unusual to say that $\sqrt{n}$ has degree $1/2$, but we will be bolder still and say that $\sqrt{n^3+1}$ has degree $3/2$, and that $\sqrt{n^3+1}/n$ has degree $3/2 - 1 = 1/2$, for example. In this way we can assign a degree to any algebraic expression. It can be shown that the degree of an expression is well-defined.

Example 1.61 *Analyze the behavior of the series*

$$\sum_{n=1}^{\infty} \frac{\sqrt{n^2+4}}{(n^{1/3}+19)^5}.$$

Solution. The numerator has degree $2 \cdot \frac{1}{2} = 1$ and the denominator has degree $\frac{1}{3} \cdot 5 = \frac{5}{3}$. The overall series, then, has degree $-2/3$. This suggests the auxiliary series

$$\sum_{n=1}^{\infty} b_n = \sum_{n=1}^{\infty} \frac{1}{n^{2/3}},$$

which is a divergent p-series. Expecting to use the Limit Comparison Test, we write

$$\lim_{n\to\infty} \frac{\frac{\sqrt{n^2+4}}{(n^{1/3}+19)^5}}{\frac{1}{n^{2/3}}} = \lim_{n\to\infty} \frac{\sqrt{n^2+4}n^{-1}}{(n^{1/3}+19)^5 n^{-5/3}} = \lim_{n\to\infty} \frac{\sqrt{1+4n^{-2}}}{(1+19n^{-1/3})} = 1,$$

and we see that the given series also diverges.

1.4.5 Ratio Test

The following test is perhaps the most memorable of the basic set of convergence tests, because it is intrinsic and easy to apply. It draws on the power we established with Theorem 1.39.

Theorem 1.62 (Ratio Test) *Consider a strictly positive series $\sum_{n=0}^{\infty} a_n$, and define*

$$r = \lim_{n\to\infty} \frac{a_{n+1}}{a_n},$$

assuming that it exists.

1. *If $r < 1$, then $\sum_{n=0}^{\infty} a_n$ converges.*
2. *If $r > 1$, then $\sum_{n=0}^{\infty} a_n$ diverges.*
3. *If $r = 1$, then the test is inconclusive.*

Proof.

Case 1. $r < 1$. Choose some r' such that $r < r' < 1$. By the definition of limit and of r, we know that $a_{n+1}/a_n < r'$ for all n sufficiently large. For these large n, then, $a_{n+1}/a_n < (r')^{n+1}/r'^n$. Now the series $\sum_{n=0}^{\infty}(r')^n$ is a convergent geometric series, so we can apply the Ratio Comparison Theorem to the tail of our series to show convergence.

Case 2. $r > 1$. Choose some r' such that $r > r' > 1$. By the definition of limit and of r, we know that $a_{n+1}/a_n > r'$ for all n sufficiently large. For these large n, then, $a_{n+1}/a_n > (r')^{n+1}/(r')^n$. Now the series $\sum_{n=0}^{\infty}(r')^n$ is a divergent geometric series, so we can apply the Ratio Comparison Theorem to the tail of our series to show divergence.

Case 3. $r = 1$. Consider the series $\sum 1/n$ and $\sum 1/n^2$. As p-series, the first diverges and the second converges, but both have $r = 1$.

Theorem 1.63 (Ratio Test Strengthened) *Consider a strictly positive series $\sum_{n=0}^{\infty} a_n$, and define*

$$r_1 = \liminf_{n\to\infty} \frac{a_{n+1}}{a_n}, \qquad r_2 = \limsup_{n\to\infty} \frac{a_{n+1}}{a_n}.$$

1. *If $r_2 < 1$, then $\sum_{n=0}^{\infty} a_n$ converges.*
2. *If $r_1 > 1$, then $\sum_{n=0}^{\infty} a_n$ diverges.*

Example 1.64 (Example 1.43 Revisited Again) *Can we use the Ratio Test to analyze the behavior of the series*

$$1 + \frac{1}{\sqrt{2}} + \frac{1}{\sqrt{3}} + \frac{1}{\sqrt{4}} + \frac{1}{\sqrt{5}} + \cdots = \sum_{n=1}^{\infty} \frac{1}{\sqrt{n}}?$$

Solution. If we tried to apply the Ratio Test, we would find

$$r = \lim_{n\to\infty} \frac{\sqrt{n}}{\sqrt{n+1}} = 1,$$

so that neither the Ratio Test nor the Strengthened Ratio Test can resolve the series. Notice that this series *does* satisfy

$$\frac{a_{n+1}}{a_n} < 1$$

for all n, but this is *not* sufficient to establish convergence. In fact we already know that this series *diverges*. This is a special case of a more general phenomenon. Whenever the terms of an infinite series are given by an algebraic function, the Ratio Test will always fail with $r = 1$; the test of the next section will be similarly useless. This should not discourage the reader because such series can always be handled by the Limit Comparison Test and p-series, as mentioned earlier.

1.4.6 Root Test

The final test of this section is also an intrinsic test. While it is sometimes algebraically more difficult to apply than the preceding test, the reverse is also true, and we will see in Chapter 4 that this test is actually stronger than the Ratio Test. Notice that we are again exploiting the predictable behavior of the geometric series.

Theorem 1.65 (Root Test) *Consider a positive series $\sum_{n=0}^{\infty} a_n$, and define*

$$\rho = \lim_{n\to\infty} a_n^{1/n}$$

if it exists.

1. *If $\rho < 1$, then $\sum_{n=0}^{\infty} a_n$ converges.*
2. *If $\rho > 1$, then $\sum_{n=0}^{\infty} a_n$ diverges.*
3. *If $\rho = 1$, then the test is inconclusive.*

Proof.

Case 1. $\rho < 1$. Choose some $\rho < \rho' < 1$. By the definition of limit and of ρ, we know that $a_n^{1/n} < \rho'$ for all n sufficiently large. For these large n, then, $a_n < (\rho')^n$. Now the series $\sum_{n=0}^{\infty} (\rho')^n$ is a convergent geometric series, so we can apply the Comparison Test to the tail of our series to show convergence.

Case 2. $\rho > 1$. By the definition of limit and of ρ, we know that $a_n^{1/n} > 1$ for all n sufficiently large. For these large n, then, $a_n > 1$, so we cannot have $\lim_{n\to\infty} a_n = 0$. Then $\sum_{n=0}^{\infty} a_n$ diverges by the Divergence Test.

Case 3. $\rho = 1$. Consider the series $\sum 1/n$ and $\sum 1/n^2$. As p-series, the first diverges and the second converges, but both have $\rho = 1$.

Example 1.66 *Investigate the behavior of the series*

$$\sum_{n=1}^{\infty} \left(\frac{2n+3}{5n-4} \right)^n .$$

Solution. The presence of an nth power in the term a_n suggests using the Root Test. We write

$$\rho = \lim_{n\to\infty} \sqrt[n]{\left(\frac{2n+3}{5n-4} \right)^n} = \lim_{n\to\infty} \frac{2n+3}{5n-4} = \frac{2}{5} < 1,$$

and we have immediately that the series converges, by the Root Test.

Theorem 1.67 (Root Test Strengthened) *Consider a positive series $\sum_{n=0}^{\infty} a_n$, and define*

$$\rho_1 = \liminf_{n\to\infty} a_n^{1/n}, \qquad \rho_2 = \limsup_{n\to\infty} a_n^{1/n}.$$

1. *If $\rho_2 < 1$, then $\sum_{n=0}^{\infty} a_n$ converges.*
2. *If $\rho_1 > 1$, then $\sum_{n=0}^{\infty} a_n$ diverges.*

Example 1.68 *Investigate the behavior of the series*

$$\sum_{n=1}^{\infty} \left(1 - \frac{1}{n} \right)^n .$$

Solution. The Root Test appears to be a good choice as it was in Example 1.66. If we try, we find

$$\rho = \lim_{n\to\infty} \sqrt[n]{\left(1 - \frac{1}{n}\right)^n} = \lim_{n\to\infty} \left(1 - \frac{1}{n}\right) = 1,$$

so the Root Test is not immediately decisive. Because $\sqrt[n]{a_n}$ approaches 1 always from below and is therefore less than 1 for every n, we may well suspect that this series will converge for some reason similar to the proof of the Root Test. However, our intuition would be wrong in this. Notice that a familiar limit from calculus gives

$$\lim_{n\to\infty} \left(1 - \frac{1}{n}\right)^n = e^{-1} \neq 0,$$

so that this series actually fails even the crudest convergence criterion and diverges by the Divergence Test.

1.5 General Series

1.5.1 Absolute and Conditional Convergence

All of the preceding theorems are applicable to positive series or strictly positive series, so at present we know a fair bit about such series, but virtually nothing about general series. The desire to apply the tests we have learned in more general circumstances motivates considering the (positive) series we obtain by taking the absolute value of each term of a series.

Definition 1.69 *A series $\sum_{n=0}^{\infty} a_n$ converges absolutely if and only if $\sum_{n=0}^{\infty} |a_n|$ converges. A series that converges absolutely is often called absolutely convergent.*

Under this definition, all our theorems from the previous section become tests for absolute convergence of general series. Of course, there is no *a priori* reason to suppose that the behavior of the series of absolute values has any particular relationship to the original series. The next theorem shows that in fact they are related.

Theorem 1.70 *A series that converges absolutely converges in the ordinary sense.*

Proof. Let $\sum_{n=0}^{\infty} a_n$ be an absolutely convergent series. Consider the series

$$\textstyle\sum_{n=0}^{\infty} b_n = \sum_{n=0}^{\infty} (a_n + |a_n|).$$

Since $a_n + |a_n|$ is either 0 or $2|a_n|$ according as a_n is negative or positive, $\sum_{n=0}^{\infty} b_n$ is a positive series. Using this information,

$$\sum_{n=0}^{\infty} b_n \leq \sum_{n=0}^{\infty} 2|a_n| = 2\sum_{n=0}^{\infty} |a_n|.$$

Now, the right-hand series converges by hypothesis, so $\sum_{n=0}^{\infty} b_n$ converges by the Comparison Test. Finally, the original series $\sum_{n=0}^{\infty} a_n$ converges because it is the difference of the convergent series $\sum_{n=0}^{\infty} |a_n|$ and $\sum_{n=0}^{\infty} b_n$.

Before moving on, we wish to emphasize the power of the algebraic trick used in the problem, that of studying a_n by means of $|a_n|$ and $a_n + |a_n|$, which of course are always positive quantities. This observation allows us to offer our next result, Theorem 1.71.

Theorem 1.71 *Let $\sum_{n=0}^{\infty} a_n$ be an absolutely convergent series and let $\sum_{n=0}^{\infty} b_n$ be a rearrangement of the first series. Then the two series both converge to the same sum.*

Proof. Notice that the series $\sum_{n=0}^{\infty} |b_n|$ and $\sum_{n=0}^{\infty}(b_n + |b_n|)$ are rearrangements of $\sum_{n=0}^{\infty} |a_n|$ and $\sum_{n=0}^{\infty}(a_n + |a_n|)$. Since $\sum a_n$ is absolutely convergent, these latter two series converge. Using Theorem 1.48 and addition properties of series, we have

$$\sum_{n=0}^{\infty} b_n = \sum_{n=0}^{\infty}(b_n + |b_n|) - \sum_{n=0}^{\infty} |b_n| = \sum_{n=0}^{\infty}(a_n + |a_n|) - \sum_{n=0}^{\infty} |a_n| = \sum_{n=0}^{\infty} a_n.$$

Example 1.72 *Investigate the behavior of the series*

$$\sum_{n=1}^{\infty}(-1)^{n+1}\frac{\sin n}{n^2}.$$

Solution. This is not a positive series, because the numerator changes sign from positive to negative in an erratic sort of way. Instead we consider the associated positive term series:

$$\sum_{n=1}^{\infty}\left|(-1)^{n+1}\frac{\sin n}{n^2}\right| = \sum_{n=1}^{\infty}\frac{|\sin n|}{n^2} \le \sum_{n=1}^{\infty}\frac{1}{n^2}.$$

Since the right-hand series is a convergent p-series, we can use the Comparison Test to show that the series of absolute values is convergent. We conclude that the original series is *absolutely* convergent, hence convergent in the ordinary sense by Theorem 1.70.

The optimistic reader may be hoping that the converse will also hold, that we can understand all series by understanding positive series. Unfortunately this is not the case, and we will see a counterexample in the next section. Series that converge without doing so absolutely have a special name, and we will see later on that they have special properties. In Chapter 3, we will find out that Theorem 1.71 fails to extend to this class of series, and in fact fails in a very dramatic way.

Definition 1.73 *A series $\sum_{n=0}^{\infty} a_n$ converges conditionally if it converges but does not converge absolutely.*

1.5.2 Alternating Series

The reader may be wondering how we could ever know that a given series was conditionally convergent, if all of the tests we have devised for convergence apply only to positive series and are therefore tests for absolute convergence. In this section we give a handy and easy-to-apply test for diagnosing certain kinds of conditionally convergent series.

Definition 1.74 *An alternating series is a series whose terms alternate in sign—it can be written in one of the forms* $\sum_{n=0}^{\infty}(-1)^n a_n$ *or* $\sum_{n=0}^{\infty}(-1)^{n+1} a_n$*, where* $a_n \geq 0$.

Theorem 1.75 (Alternating Series Test) *Consider an alternating series*

$$\sum_{n=0}^{\infty}(-1)^n a_n,$$

and suppose that the terms a_n *satisfy:*

1. $\{a_n\}$ *is nonincreasing eventually; i.e., for all* $n \geq N$ *for some integer* $N > 0$ *we have* $a_{n+1} \leq a_n$.
2. $\lim_{n\to\infty} a_n = 0$.

Then $\sum_{n=0}^{\infty}(-1)^n a_n$ *converges to a sum S. Furthermore, assuming* $k \geq N$ *for some integer* $N > 0$*, we have*

$$S_k \leq S \leq S_{k+1} \quad \text{or} \quad S_{k+1} \leq S \leq S_k$$

according to whether k is odd or even. (Consequently, S and S_k *can never differ by more than* a_{k+1}*.)*

Proof. Because we are concerned only with the behavior of the tail of any series, we lose no generality in supposing that $a_{n+1} \leq a_n$ holds universally. As usual let $\{s_k\}$ be the sequence of partial sums. Consider for the moment just the odd indexed sums $\{s_{2k+1}\}$, and write

$$s_{2k+1} = (a_0 - a_1) + (a_2 - a_3) + (a_4 - a_5) + \cdots + (a_{2k+1}).$$

We can view the s_{2k+1} as the sequence of partial sums of a positive series, since the terms $a_n - a_{n+1}$ are nonnegative. On the other hand,

$$s_{2k+1} = a_0 - (a_1 - a_2) - (a_3 - a_4) - \cdots - a_{2k+1} < a_0.$$

Since $\{s_{2k+1}\}$ is monotone increasing and bounded above it must converge to a real number S. Furthermore,

$$\lim_{k\to\infty} s_{2k} = \lim_{k\to\infty}(s_{2k+1} - a_{2k+1}) = S - 0 = S.$$

Since both even- and odd-indexed partial sums converge to S, the series converges.

This quick argument demonstrates the convergence of the alternating series. To prove the latter half of the theorem, we will interpret the behavior of an alternating series in a more visual way (we encourage the reader to draw the picture based on this chain of inequalities). We have already shown

$$s_1 \leq s_3 \leq s_5 \leq \cdots,$$

and if we look at the even partial sums in a similar way, we have

$$s_1 \le s_3 \le s_5 \le \cdots \le s_{2k+1} \le S \le s_{2k} \le \cdots \le s_4 \le s_2 \le s_0.$$

Example 1.76 *Investigate the behavior of the series*

$$\frac{1}{4} - \frac{1}{7} + \frac{1}{12} - \frac{1}{19} + \cdots = \sum_{n=1}^{\infty} (-1)^{n+1} \frac{1}{n^2+3}.$$

Solution. While the reader can verify that the Alternating Series Test does apply, something stronger is true. We test for absolute convergence by considering the associated positive term series

$$\sum_{n=1}^{\infty} \left| (-1)^{n+1} \frac{1}{n^2+3} \right| = \sum_{n=1}^{\infty} \frac{1}{n^2+3}.$$

However, in this form we obtain

$$\frac{1}{n^2+3} \le 1/n^2.$$

Since

$$\sum_{n=1}^{\infty} 1/n^2$$

is a convergent p-series, the basic Comparison Test shows that the associated positive term series is convergent. That is,

$$\sum_{n=1}^{\infty} (-1)^{n+1} \frac{1}{n^2+3}$$

is absolutely convergent (and hence convergent).

Example 1.77 *Investigate the behavior of the series*

$$\frac{1}{2} - \frac{2}{5} + \frac{3}{10} - \frac{4}{17} + \cdots = \sum_{n=1}^{\infty} (-1)^{n+1} \frac{n}{n^2+1}.$$

Solution. Unlike the previous example, we cannot establish convergence by considering the associated positive term series. Notice that the series has the form

$$\sum_{n=1}^{\infty} \left| (-1)^{n+1} \frac{n}{n^2+1} \right| = \sum_{n=1}^{\infty} \frac{n}{n^2+1}.$$

By using (for example) the Limit Comparison Test with the *divergent* harmonic series $\sum_{n=1}^{\infty} 1/n$, we see that the original series fails to converge absolutely.

In this case we really do use the Alternating Series Test. There are two things to check.

$$\begin{aligned} a_{n+1} - a_n & \\ &= \frac{n+1}{(n+1)^2+1} - \frac{n}{n^2+1} \\ &= \frac{(n+1)(n^2+1) - n(n^2+2n+2)}{(n^2+1)(n^2+2n+2)} \\ &= \frac{-n^2-n+1}{(n^2+1)(n^2+2n+2)}. \end{aligned}$$

Since this is negative when $n \geq 1$, the terms are decreasing in absolute value. Since in this case the terms

$$a_n = \frac{n}{n^2+1}$$

are given as a differentiable function of n, we can also check that they are decreasing by taking the derivative. Here, if we write

$$a(x) = \frac{x}{x^2+1},$$

we have

$$a'(x) = \frac{1(x^2+1) - 2x(x)}{(x^2+1)^2} = \frac{-x^2}{(x^2+1)^2} < 0.$$

Also,

$$\lim_{n\to\infty} \frac{n}{n^2+1} = 0,$$

as desired.

Thus the conditions for the Alternating Series Test are met, and the given series converges. Since the series failed to converge absolutely and does converge as given, it converges conditionally.

Example 1.78 *Investigate the behavior of the series*

$$\frac{2}{3} - \frac{3}{5} + \frac{4}{7} - \frac{5}{9} + \cdots = \sum_{n=2}^{\infty} (-1)^n \frac{n}{2n-1}.$$

Solution. Recognizing this as an alternating series, we try to apply the Alternating Series Test. We have

$$a_{n+1} - a_n = \frac{n+1}{2n+1} - \frac{n}{2n-1} = \frac{(n+1)(2n-1) - n(2n+1)}{(2n+1)(2n-1)} = \frac{-1}{4n^2-1} < 0,$$

so that the terms are decreasing in absolute value. However,

$$\lim_{n\to\infty} \frac{n}{2n-1} = \frac{1}{2} \neq 0,$$

so the terms do not converge to 0, hence the Alternating Series does not guarantee convergence. In fact, this forces the series to *diverge* by the Divergence Test.

Due to the power of the Divergence Test to be decisive very quickly, it is generally best to check $\lim_{n\to\infty} a_n$ first to see if the answer is zero, and only then to check if

$$a_{n+1} - a_n \leq 0.$$

2

More Sophisticated Techniques

It is an unfortunate truth that most students of calculus who encounter infinite series are given only a small collection of theorems about the convergence and divergence of infinite series. Students typically learn the Divergence Test, the Alternating Series Test, the Ratio Test, the Root Test, and some version of the Comparison Tests. Generally the message is that beyond these preliminary results the study of infinite series becomes very hard very fast. However, there are some easily constructed series that do not yield to any of the tests just listed. It is the learned response of calculus students to throw up their hands when the Ratio Test and Root Test give the inconclusive limit of 1. The authors certainly acknowledge that there does not exist a test or set of tests that can systematically determine whether every infinite series converges or diverges. However, we believe strongly that there are several more sophisticated and more powerful tests that are just as easy to use and that can decide the convergence of nearly all series that arise in practice. For whatever reason, these tests are not well known except among specialists of infinite series.

In this short chapter we include a small collection of more sophisticated tests. While some of these tests are slightly more difficult to apply than the tests of Chapter 1, these tests are generally both easy to use and powerful. Series that defy all of these tests do exist, but are generally quite pathological. It is our hope that the reader will no longer be in a position of despair when confronting a series for which the Root Test fails.

2.1 The Work of Cauchy

Theorem 2.1 (Cauchy Criterion) *Let $\sum_{n=0}^{\infty} a_n$ be an infinite series. Then $\sum_{n=0}^{\infty} a_n$ converges if and only if, for every $\epsilon > 0$, there exists a positive integer N with the property that, for every $n_1 \geq n_0 \geq N$,*

$$\left| \sum_{n=n_0}^{n_1} a_n \right| < \epsilon.$$

A series that satisfies this criterion is said to be a Cauchy series.

Proof. As always, we write $S_k = \sum_{n=0}^{k} a_n$ for the partial sums of the series. Then the series converges if and only if $\{S_k\}$ converges.

This is just a restatement of Theorem 1.46, translated into the language of series. However, because we are choosing to focus on infinite series in this book, we will prove it again here.

To show the "if" case, assume that for every $\epsilon > 0$, there exists a positive integer N with the property that, for every $n_1 \geq n_0 \geq N$,

$$\left|\sum_{n=n_0}^{n_1} a_n\right| < \epsilon \Leftrightarrow \left|S_{n_1} - S_{n_0-1}\right| < \epsilon.$$

Write $L = \liminf_{n\to\infty} S_n$, $R = \limsup_{n\to\infty} S_n$. If $L = R$, then $\{S_n\}$ converges to their common value. Assume for the sake of contradiction that $R - L = E > 0$. By hypothesis, we can find an N such that for every $n_1 \geq n_0 \geq N$, $|S_{n_1} - S_{n_0-1}| < E/3$. Since L and R are limit points, we can find infinitely many indices i, j such that $|L - S_i| < E/3$, $|R - S_j| < E/3$. In particular we can choose $i, j > N + 1$, so that $|S_i - S_j| < E/3$. Then

$$R - L = |R - L| \leq |R - S_j| + |S_j - S_i| + |L - S_i| < E/3 + E/3 + E/3 = E = R - L,$$

a contradiction.

To show the "only if" case, suppose that $\{S_k\}$ converges to a limit S. Let $\epsilon > 0$ be given. Then there exists a positive integer N such that, for all $n > N$, $|S_n - S| < \epsilon/2$. Then, for all $n_1 \geq n_0 \geq N + 1$, we have

$$\left|S_{n_1} - S_{n_0-1}\right| \leq |S_{n_1} - S| + |S_{n_0-1} - S| < \epsilon.$$

Remark. The content and proof of the above theorem can be summarized as follows. A series converges if and only if the associated sequence of partial sums converges if and only if the associated sequence of partial sums is Cauchy.

There are at least two ways to use this theorem. First, sometimes it is possible to estimate $|S_m - S_n|$ directly in terms of m and n and consider the limit. This is a bit optimistic for a complicated series, however. This theorem can also be used as a test for divergence. If a given series has arbitrarily many disjoint "blocks" of consecutive terms with a large sum (say, at least ϵ for some fixed $\epsilon > 0$), it must diverge, since the Cauchy criterion will fail for that same choice of ϵ. (This can be regarded as an upgraded version of the usual Divergence Test.)

Example 2.2 *Suppose that $\sum_{n=1}^{\infty} a_n$ is a convergent series of positive terms, and let $r_n = \sum_{k=n}^{\infty} a_k$ be the nth tail for each n. Write $b_n = a_n/r_n$. Then*

$$\sum_{n=1}^{\infty} b_n$$

diverges.

Solution. We will show that the Cauchy criterion is not satisfied. In particular, for any N, we will find a block of terms $b_m + b_{m+1} + b_{m+2} + \cdots + b_n$ with $n > m > N$ such that

$b_m + b_{m+1} + b_{m+2} + \cdots + b_n > 1/2$. To this end, let N be given, and take any $m > N$. Then we have

$$r_m = a_m + a_{m+1} + a_{m+2} + \cdots = \lim_{n\to\infty} (a_m + a_{m+1} + \cdots + a_n).$$

In particular, for n large enough, we have

$$a_m + a_{m+1} + \cdots + a_n > \frac{r_m}{2}.$$

Then we can write, noticing that the sequence $\{r_n\}$ is strictly decreasing,

$$\begin{aligned} b_m + b_{m+1} + b_{m+2} + \cdots + b_n &= \frac{a_m}{r_m} + \frac{a_{m+1}}{r_{m+1}} + \cdots + \frac{a_n}{r_n} \\ &\geq \frac{a_m}{r_m} + \frac{a_{m+1}}{r_m} + \cdots + \frac{a_n}{r_m} \\ &= \frac{a_m + a_{m+1} + \cdots + a_n}{r_m} \\ &> \frac{1}{2}. \end{aligned}$$

Since the Cauchy Criterion fails, the series diverges.

Theorem 2.3 (Cauchy Condensation Test) *Let $\sum_{n=0}^{\infty} a_n$ be a positive infinite series, and suppose that the terms a_n are eventually decreasing. Then the series*

$$\sum_{n=0}^{\infty} a_n \quad \textit{and} \quad \sum_{n=0}^{\infty} 2^n a_{2^n}$$

either both converge or both diverge.

Proof. Since changing any finite number of terms in a series cannot change whether it converges or diverges, we may assume without loss of generality that the terms are decreasing throughout. Let

$$S_n = \sum_{k=0}^{n} a_k; \quad \text{and} \quad T_n = \sum_{k=0}^{n} 2^k a_{2^k}$$

be the partial sums of the series in question. Then we can write

$$\begin{aligned} S_1 &= a_0 + a_1 \\ &= a_0 + T_0 \\ S_3 &= S_1 + a_2 + a_3 \leq a_0 + T_0 + a_2 + a_2 \\ &= a_0 + T_0 + 2a_2 = a_0 + T_1 \\ S_7 &= S_3 + a_4 + a_5 + a_6 + a_7 \leq a_0 + T_1 + 4a_4 \\ &= a_0 + T_2 \end{aligned}$$

and in general

$$S_{2^n-1} \leq a_0 + T_{n-1}.$$

On the other hand,

$$\begin{aligned} S_2 &= a_0 + a_1 + a_2 \geq a_1 + a_2 + a_2 = a_1/2 + a_2 + a_1/2 + a_2 \\ &= a_1/2 + a_2 + T_1/2 \\ S_4 &= S_2 + a_3 + a_4 \geq T_1 + a_4 + a_4 = a_1/2 + a_2 + T_1/2 + 2a_4 \\ &= a_1/2 + a_2 + T_2/2 \\ S_8 &= S_4 + a_5 + a_6 + a_7 + a_8 \geq a_1/2 + a_2 + T_2/2 + 4a_8 \\ &= a_1/2 + a_2 + T_3/2 \end{aligned}$$

and in general

$$S_{2^n} \geq a_1/2 + a_2 + T_n/2.$$

Since the two series in question are positive, they cannot oscillate. That is, the partial sums will either increase to a limit or diverge to infinity. If the sequence $\{S_n\}$ increases to infinity, then the first set of inequalities show that $\{T_n\}$ does also. Similarly, the second set of inequalities show that, if $\{T_n\}$ diverges to infinity, $\{S_n\}$ does also. That is, the two series converge or diverge together.

The classic application of this result is to show that the harmonic series diverges, which is our Proof 1 in Chapter 3.

Although the appearance of 2^n in the above theorem seems reasonable in light of the proof, it is in a sense arbitrary. Would 3^n work? Would n^2? The following generalization by Oscar Xavier Schlömilch (1823–1901) shows that we can do much better.

Theorem 2.4 (Schlömilch) *Let $\sum_{n=0}^{\infty} a_n$ be a positive infinite series, and suppose that the terms a_n are eventually decreasing. Let $n_0 < n_1 < n_2 < \cdots$ be a strictly increasing sequence of positive integers such that $(n_{k+1} - n_k)/(n_k - n_{k-1})$ is bounded (as a function of k). Then the series*

$$\sum_{n=0}^{\infty} a_n \quad \text{and} \quad \sum_{k=0}^{\infty} (n_{k+1} - n_k) a_{n_k}$$

either both converge or both diverge.

Proof. Since changing any finite number of terms in a series cannot change whether it converges or diverges, we may assume without loss of generality that the terms are decreasing throughout. Let

$$S_n = \sum_{k=0}^{n} a_k$$

$$T_n = \sum_{k=0}^{n} (n_{k+1} - n_k) a_{n_k}$$

$$U_n = \sum_{k=1}^{n} (n_k - n_{k-1}) a_{n_k}$$

be the partial sums of the series in question. Then we can write

$$\begin{aligned} S_{n_1} &= S_{n_0} + a_{n_0+1} + a_{n_0+2} + \cdots + a_{n_1} \geq S_{n_0} + (n_1 - n_0)a_{n_1} \\ &= S_{n_0} + U_1 \\ S_{n_2} &= S_{n_1} + a_{n_1+1} + a_{n_1+2} + \cdots + a_{n_2} \geq S_{n_0} + U_1 + (n_2 - n_1)a_{n_2} \\ &= S_{n_0} + U_2 \end{aligned}$$

and in general

$$S_{n_k} \geq S_{n_0} + U_k.$$

On the other hand,

$$\begin{aligned} S_{n_1-1} &= S_{n_0-1} + a_{n_0} + a_{n_0+1} + \cdots + a_{n_1-1} \leq S_{n_0-1} + (n_1 - n_0)a_{n_0} \\ &= S_{n_0-1} + T_0 \\ S_{n_2-1} &= S_{n_1-1} + a_{n_1} + a_{n_1+1} + \cdots + a_{n_2-1} \leq S_{n_0-1} + T_0 + (n_2 - n_1)a_{n_1} \\ &= S_{n_0-1} + T_1 \end{aligned}$$

and in general

$$S_{n_k-1} \leq S_{n_0-1} + T_{k-1}.$$

Arguing as in the previous theorem, we see that, if $\{S_n\}$ diverges, $\{U_n\}$ diverges also, and further that if $\{T_n\}$ diverges, $\{S_n\}$ diverges. Then the series $\sum_{n=0}^{\infty} a_n$ converges if $\sum_{k=1}^{\infty} (n_k - n_{k-1}) a_{n_k}$ converges and only if $\sum_{k=0}^{\infty} (n_{k+1} - n_k) a_{n_k}$ converges. However, since $(n_{k+1} - n_k)/(n_k - n_{k-1})$ is bounded, we can apply the Strengthened Limit Comparison Test to the latter two series and conclude that they converge or diverge together. That is, all three series have the same limit behavior, as desired.

Example 2.5 *Investigate the convergence of*

$$\sum_{n=1}^{\infty} \frac{1}{2^{\sqrt{n}}}$$

Solution. Though this series resembles a geometric series, the presence of the square root complicates matters. If we attempt to apply the Ratio Test, we find much to our chagrin that

$$\lim_{n\to\infty} \frac{\frac{1}{2^{\sqrt{n+1}}}}{\frac{1}{2^{\sqrt{n}}}} = \lim_{n\to\infty} 2^{\sqrt{n}-\sqrt{n+1}} = 2^{\lim_{n\to\infty}(\sqrt{n}-\sqrt{n+1})} = 2^0 = 1,$$

giving us no information at all. The Root Test is similarly inconclusive.

The correct test is the Schlömilch Condensation Test. Let $n_k = k^2$, and notice that indeed $[(k+1)^2 - k^2]/[k^2 - (k-1)^2] = (2k+1)/(2k-1) \leq 3$ is bounded. Since the terms are positive and decreasing (check this), the test applies, and we can say that

$$\sum_{n=1}^{\infty} \frac{1}{2^{\sqrt{n}}}$$

converges if and only if

$$\sum_{k=1}^{\infty} \frac{(k+1)^2 - k^2}{2^{\sqrt{k^2}}} = \sum_{k=1}^{\infty} \frac{2k+1}{2^k}$$

converges. However, now the Ratio Test will succeed because we have removed the square root.

$$\lim_{n\to\infty} \frac{\frac{2(k+1)+1}{2^{k+1}}}{\frac{2k+1}{2^k}} = \lim_{n\to\infty} \frac{2k+3}{4k+2} = 1/2 < 1,$$

and the series converges.

2.2 Kummer's Results

In this section we give a strong criterion for convergence. This criterion comes essentially from a clever algebraic trick, which is due to Eduard Kummer (1810–1893).

Theorem 2.6 *Let $\sum_{n=0}^{\infty} a_n$ be a positive infinite series, and let $\{\lambda_n\}$ be any sequence of positive numbers. If, eventually,*

$$\lambda_n - \frac{a_{n+1}}{a_n}\lambda_{n+1} \geq k > 0$$

for some constant k, then $\sum_{n=0}^{\infty} a_n$ converges.

Proof. As usual, we can assume without loss of generality that the eventual condition in fact holds everywhere.

Rearranging the given inequality, we have

$$\frac{1}{k}(a_n\lambda_n - a_{n+1}\lambda_{n+1}) \geq a_n.$$

Summing this inequality for $n = 0, 1, \ldots, m$, we obtain

$$\sum_{n=0}^{m} a_n \leq \frac{1}{k}(a_0\lambda_0 - a_{m+1}\lambda_{m+1}) \leq \frac{1}{k}a_0\lambda_0.$$

That is, the partial sums of $\sum_{n=0}^{\infty} a_n$ are bounded above. Since this series is positive, its partial sums are monotone increasing, and this suffices to show convergence.

Theorem 2.7 (Kummer's Test) *Let $\sum_{n=0}^{\infty} a_n$ be a positive infinite series, let $\sum_{n=0}^{\infty} 1/d_n$ be a divergent positive series, and define*

$$\kappa_n = d_n - \frac{a_{n+1}}{a_n} d_{n+1}.$$

1. *If, eventually, $\kappa_n > k > 0$ for some constant k, then $\sum_{n=0}^{\infty} a_n$ converges.*
2. *If, eventually, $\kappa_n \leq 0$, then $\sum_{n=0}^{\infty} a_n$ diverges.*

Proof. Without loss of generality we drop the word "eventually" from the conditions given. The first half of the theorem merely restates the preceding theorem, so we consider the second half. If

$$d_n - \frac{a_{n+1}}{a_n} d_{n+1} \leq 0,$$

we can rearrange this to

$$\frac{a_{n+1}}{a_n} \geq \frac{1/d_{n+1}}{1/d_n},$$

and the divergence of $\sum_{n=0}^{\infty} a_n$ now follows from the Ratio Comparison Test.

Corollary 2.7.1 (Kummer's Test Simplified) *Let $\sum_{n=0}^{\infty} a_n$ be a positive infinite series, and let $\sum_{n=0}^{\infty} 1/d_n$ be a divergent positive series, and define*

$$\kappa = \lim_{n\to\infty} \left(d_n - \frac{a_{n+1}}{a_n} d_{n+1} \right)$$

assuming that it exists. If $\kappa > 0$ (resp. $\kappa < 0$), then $\sum_{n=0}^{\infty} a_n$ converges (resp. diverges).

Proof. We simply notice that, if

$$\lim_{n\to\infty} \left(d_n - \frac{a_{n+1}}{a_n} d_{n+1} \right) = \kappa > 0,$$

we must eventually have

$$d_n - \frac{a_{n+1}}{a_n} d_{n+1} > \frac{\kappa}{2}.$$

Similarly, if

$$\lim_{n\to\infty}\left(d_n - \frac{a_{n+1}}{a_n}d_{n+1}\right) = \kappa < 0,$$

we must eventually have

$$d_n - \frac{a_{n+1}}{a_n}d_{n+1} < \frac{\kappa}{2}.$$

In either case, we need only apply the preceding theorem.

Kummer's tests are valuable because they are extremely versatile. Indeed it is not difficult to show that every series can be shown to converge or diverge for a suitable choice of the d_n. It is precisely this statement of power that suggests the downfall of Kummer's tests: they are fundamentally extrinsic. The choice of d_n has to come from the solver, and usually there is no particular motivation.

However, Kummer's tests are of high practical value in proving some important intrinsic tests, viz. the ones in the coming section.

2.3 The Tests of Raabe and Gauss

The following test is due to the German mathematician and physicist Joseph Ludwig Raabe (1801–1859), who taught for much of his life at the Institute of Zurich. An essentially equivalent test was discovered (more or less simultaneously) by Farkas Bolyai (sometimes known as Wolfgang). Bolyai is principally remembered not for his inquiry into infinite series but for his study of geometry, on which he spent most of his professional time.

Theorem 2.8 (Raabe's Test) *Let $\sum_{n=0}^{\infty} a_n$ be a positive infinite series, and suppose we can write*

$$\frac{a_{n+1}}{a_n} = 1 - \frac{\beta_n}{n},$$

where

$$\lim_{n\to\infty} \beta_n = \beta.$$

1. *If $\beta > 1$, then $\sum_{n=0}^{\infty} a_n$ converges.*
2. *If $\beta < 1$, then $\sum_{n=0}^{\infty} a_n$ diverges.*

Proof. We first rewrite

$$\frac{a_{n+1}}{a_n} = 1 - \frac{\beta_n}{n} = \frac{n-1}{n} - \frac{\beta_n - 1}{n}$$

as

$$(n-1) - \frac{a_{n+1}}{a_n}n = \beta_n - 1.$$

We can now take $d_n = n - 1$, noting that the series $\sum_{n=2}^{\infty} 1/d_n$ is the harmonic series, divergent by Theorem 1.59. The result is now seen as a special case of Kummer's Test.

Theorem 2.9 (Raabe's Test Strengthened) *Let $\sum_{n=0}^{\infty} a_n$ be a positive infinite series, and suppose we can write*

$$\frac{a_{n+1}}{a_n} = 1 - \frac{\beta_n}{n},$$

where

$$\limsup_{n\to\infty} \beta_n = \beta_{\uparrow}; \ \liminf_{n\to\infty} \beta_n = \beta_{\downarrow}.$$

1. *If $\beta_{\downarrow} > 1$, then $\sum_{n=0}^{\infty} a_n$ converges.*
2. *If $\beta_{\uparrow} < 1$, then $\sum_{n=0}^{\infty} a_n$ diverges.*

Example 2.10 *Show that the series*

$$\sum_{n=1}^{\infty} \frac{1 \cdot 3 \cdot 5 \cdots (2n-1)}{2 \cdot 4 \cdot 6 \cdots (2n)}$$

diverges.

Solution. Because the ratios of consecutive terms are easy to compute, it is natural to try the Ratio Test. However, it will fail here. We use Raabe's Test. Writing

$$a_n = \frac{1 \cdot 3 \cdot 5 \cdots (2n-1)}{2 \cdot 4 \cdot 6 \cdots (2n)},$$

we have

$$\frac{a_{n+1}}{a_n} = \frac{2n+1}{2n+2} = 1 - \frac{1}{2n+2} = 1 - \frac{n/(2n+2)}{n},$$

so that

$$\beta_n = \frac{n}{2n+2} \to \frac{1}{2},$$

which is less than 1. The series diverges.

Example 2.11 *Show that the series*

$$\sum_{n=1}^{\infty} \frac{1 \cdot 3 \cdot 5 \cdots (2n-1)}{2 \cdot 4 \cdot 6 \cdots (2n)(n+1)}$$

converges.

Solution. Again, the Ratio Test is not delicate enough for these series. We use Raabe's Test. Writing

$$a_n = \frac{1 \cdot 3 \cdot 5 \cdots (2n-1)}{2 \cdot 4 \cdot 6 \cdots (2n)(n+1)},$$

we have

$$\frac{a_{n+1}}{a_n} = \frac{(2n+1)(n+2)}{(2n+2)(n+1)} = \frac{2n^2+5n+2}{2n^2+4n+2} = 1 + \frac{n}{2n^2+4n+2}$$
$$= 1 - \frac{-n^2/(2n^2+4n+2)}{n},$$

so that

$$\beta_n = \frac{-n^2}{2n^2+4n+2} \to \frac{-1}{2},$$

which is less than 1. The series converges.

A stronger test that copes with the $\beta = 1$ case of Raabe's Test follows. This test is due to the prolific mathematician Karl Friedrich Gauss (1777–1855).

Theorem 2.12 (Gauss's Test) *Let $\sum_{n=0}^{\infty} a_n$ be a positive infinite series, and suppose we can write*

$$\frac{a_{n+1}}{a_n} = 1 - \frac{\beta}{n} + \frac{\theta_n}{n^{1+k}},$$

where $\{\theta_n\}$ is a bounded sequence and $k > 0$.

1. *If $\beta > 1$, then $\sum_{n=0}^{\infty} a_n$ converges.*
2. *If $\beta \leq 1$, then $\sum_{n=0}^{\infty} a_n$ diverges.*

Proof. In the case $\beta \neq 1$, we can write

$$\frac{a_{n+1}}{a_n} = 1 - \frac{\beta_n}{n},$$

where

$$\beta_n = \beta + \theta_n n^{-k} \to \beta,$$

so that we can apply Raabe's Test directly.

Assume then that $\beta = 1$, so that

$$\frac{a_{n+1}}{a_n} = 1 - \frac{1}{n} + \frac{\theta_n}{n^{1+k}}.$$

We will take $\sum_{n=2}^{\infty} 1/d_n = \sum_{n=2}^{\infty} 1/n \log n$, which diverges by the Integral Test. With that in mind we rewrite the equation from Kummer's Test as

$$(n-1)\log n - n \log n \frac{a_{n+1}}{a_n} = \frac{-\theta_n \log n}{n^k}$$

and then

$$(n-1)\log(n-1) - n\log n\frac{a_{n+1}}{a_n} = \frac{\theta_n \log n}{n^k} + (n-1)(\log(n-1) - \log n) = \kappa_n.$$

To apply Kummer's Test, we must take the limit of the right-hand side. Using L'Hopital's rule, we have

$$\lim_{n\to\infty} \frac{\log n}{n^k} = \lim_{n\to\infty} \frac{1/n}{kn^{k-1}} = \lim_{n\to\infty} \frac{1}{kn^k} = 0$$

and

$$\begin{aligned}\lim_{n\to\infty}(n-1)(\log(n-1) - \log n) &= \lim_{n\to\infty} \frac{(\log(n-1) - \log n)}{1/(n-1)} \\ &= \lim_{n\to\infty} \frac{1/(n-1) - 1/n}{1/(n-1)^2} \\ &= \lim_{n\to\infty} -\frac{n-1}{n} \\ &= -1.\end{aligned}$$

Since the θ_n are bounded, the first term in κ_n approaches 0. Since the second term approaches -1, the divergent case of Kummer's Test applies.

Example 2.13 *Let $p > 0$. Show that the series*

$$\sum_{n=1}^{\infty} \frac{p(p+1)(p+2)\cdots(p+n-1)}{n!}$$

diverges.

Proof. We use Raabe's Test. Writing

$$a_n = \frac{p(p+1)(p+2)\cdots(p+n-1)}{n!},$$

we compute the ratio

$$\frac{a_{n+1}}{a_n} = \frac{p+n}{n+1} = 1 - \frac{1-p}{n+1} = 1 - \frac{(1-p)n/(n+1)}{n}$$

so that

$$\beta_n = \frac{(1-p)n}{n+1} \to 1 - p,$$

which is less than 1. The series diverges.

While the above series yielded to Raabe's Test, the following generalization requires Gauss's sharper test.

Example 2.14 *Let $a, b > 0$. Show that the series*

$$\sum_{n=1}^{\infty} \frac{a(a+1)(a+2)\cdots(a+n-1)}{b(b+1)(b+2)\cdots(b+n-1)}$$

converges if $b-a > 1$ and diverges if $b-a \leq 1$.

Proof. First, we might try Raabe's Test; though this test would handle the cases $b-a > 1$, $b-a < 1$, it would not handle the boundary case. Writing

$$a_n = \frac{a(a+1)(a+2)\cdots(a+n-1)}{b(b+1)(b+2)\cdots(b+n-1)},$$

we compute the ratio

$$\begin{aligned}\frac{a_{n+1}}{a_n} &= \frac{a+n}{b+n} = 1 - \frac{b-a}{n+b} = 1 - \frac{(b-a)n/(n+b)}{n} \\ &= 1 - \frac{b-a}{n} + \frac{(b-a)b/(n+b)}{n} = 1 - \frac{b-a}{n} + \frac{(b-a)bn/(n+b)}{n^2}.\end{aligned}$$

so that, since the numbers $(b-a)bn/(n+b)$ are bounded, we have the kind of expression we need for Gauss's Test. Then the series converges if and only if $b-a > 1$.

2.4 Logarithmic Scales

In Chapter 1, we proved a variety of comparison tests in different forms. Extrinsic tests such as the comparison tests are only as valuable as the user's personal collection of known series. Here we give an important family of infinite series; many naturally arising series can be classified by comparing them with a series in the logarithmic scale. This scale of tests is named for the Norwegian mathematician Niels Abel (1802–1829) who gave his name to abelian groups, and the lesser-known Italian mathematician Ulisse Dini (1845–1918).

Notation 2.15 *Define $\log_{(0)} x = x$, $\log_{(1)} x = \log x$, $\log_{(2)} x = \log(\log x)$, and, inductively, $\log_{(n+1)} x = \log(\log_{(n)} x)$. Here it is understood that these iterated logarithms are defined only eventually.*

We first emphatically state that this notation conflicts with the usual usage of $\log_b$ to mean logarithm to the base b. However, we will never use logarithms to bases other than e, so we find this notation convenient.

Theorem 2.16 (Abel-Dini Scale) *For each integer k, the series*

$$\sum_{n=N}^{\infty} \frac{1}{n \log_{(1)} n \log_{(2)} n \cdots \log_{(k)} n (\log_{(k+1)} n)^{\alpha}}$$

converges if $\alpha > 1$ and diverges if $\alpha \leq 1$, where N is taken large enough for the iterated logarithms to be defined.

Proof. If $k = 0$ and $\alpha < 0$, then the terms of the series tend to ∞ thus the series diverges. Otherwise, the terms are positive and eventually monotonically decreasing, so that Cauchy's Integral Test applies.

Letting $u = \log_{(k+1)}(x)$,

$$\int_N^\infty \frac{dx}{x \log_{(1)} x \log_{(2)} x \cdots \log_{(k)} x (\log_{(k+1)}(x))^\alpha}$$

$$= \int_N^\infty \frac{du}{u^\alpha} = \begin{cases} (1-\alpha)^{-1} u^{1-\alpha}]_N^\infty = \infty & \text{if } \alpha < 1 \\ \log u]_N^\infty = \infty & \text{if } \alpha = 1 \\ (1-\alpha)^{-1} u^{1-\alpha}]_N^\infty < \infty & \text{if } \alpha > 1 \end{cases}$$

by basic limit properties, and the proposition follows.

We remark that this series in some sense generalizes our Theorem 1.59 for p-series. Like p-series, the series in the Abel-Dini Scale not only allow comparison with many naturally-arising series, but also appear directly in practical situations. For example, in our Chapter 4, we see series of this type in Gems 19, 32, and 101.

One sequence of tests that is particularly little-known was developed by Bertrand. Of the few books that even refer to Bertrand's Test, the vast majority state only one form of the tests, typically only stating the case $k = 0$ or $k = 1$. The proofs of the following three forms of Bertrand's Test are long and technical, and can be found at the end of the chapter. In essence the tests are proven using the Comparison Test using the Abel-Dini scale as the standard of reference. In a sense these tests can be considered an *intrinsic* version of the Abel-Dini scale. In some cases they are more straightforward to apply than the Abel-Dini scale because they deal directly with the ratio of the terms in the series.

Theorem 2.17 (Bertrand's Tests, First Form) *Let $\sum_n^\infty a_n$ be a series of positive real numbers. Let k be a nonnegative integer, and define a sequence $\{b_n\}$ implicitly by*

$$\frac{a_n}{a_{n+1}} = 1 + \frac{1}{n} + \frac{1}{n \log n} + \cdots$$

$$+ \frac{1}{n \log n \log_{(2)} n \cdots \log_{(k-1)} n} + \frac{b_n}{n \log n \log_{(2)} n \cdots \log_{(k-1)} n \log_{(k)} n}.$$

Then, if $\lim_{n\to\infty} b_n > 1$, $\sum_n^\infty a_n$ converges; if $\lim_{n\to\infty} b_n < 1$, $\sum_n^\infty a_n$ diverges.

More generally, if $\liminf_{n\to\infty} b_n > 1$ (resp. $\limsup_{n\to\infty} b_n < 1$), $\sum_n^\infty a_n$ converges (resp. diverges).

Theorem 2.18 (Bertrand's Tests, Second Form) *Let $\sum_n^\infty a_n$ be a series of positive real numbers. Let k be a nonnegative integer, and define a sequence $\{b_n\}$ implicitly by*

$$\frac{a_{n+1}}{a_n} = 1 - \frac{1}{n} - \frac{1}{n \log n} - \cdots$$

$$- \frac{1}{n \log n \log_{(2)} n \cdots \log_{(k-1)} n} - \frac{b_n}{\log n \log_{(2)} n \cdots \log_{(k-1)} n \log_{(k)} n}.$$

Then, if $\lim_{n\to\infty} b_n > 1$, $\sum_n^\infty a_n$ *converges; if* $\lim_{n\to\infty} b_n < 1$, $\sum_n^\infty a_n$ *diverges.*

More generally, if $\liminf_{n\to\infty} b_n > 1$ *(resp.* $\limsup_{n\to\infty} b_n < 1$*),* $\sum_n^\infty a_n$ *converges (resp. diverges).*

One very positive aspect to these tests is that they are systematic. If Bertrand's Test is inconclusive for a particular k value, there is an obvious next test to try—Bertrand's Test for $k+1$. At first glance the algebra involved in forming the relevant sequence $\{b_n\}$ for different k values is prohibitive. However, the following result shows that the work can be done by a simple algorithm that, in effect, constructs the sequences recursively, so that the algebraic effort involved in each test applies directly to subsequent tests. It is this property that gives the Bertrand tests their utility.

Theorem 2.19 (Bertrand Simplified) *Define operators* B^+, B^- *that act on sequences by*

$$B^+\{b_n\}_{n=N}^\infty = \left\{n\left(\frac{b_n}{b_{n+1}} - 1\right)\right\}_{n=N}^\infty$$

$$B^-\{b_n\}_{n=N}^\infty = \left\{n\left(1 - \frac{b_{n+1}}{b_n}\right)\right\}_{n=N}^\infty$$

For each positive integer k, let R_k *be the operator defined by*

$$R_k\{b_n\}_{n=N}^\infty = \{\log_{(k)}(n)(b_n - 1)\}_{n=N}^\infty.$$

Let $\sum_{n=N}^\infty a_n$ *be a series of positive real numbers. Let k be a nonnegative integer, and define a sequence* $\{b_n\}_{n=N}^\infty$ *by either*

$$\{b_n\}_{n=N}^\infty = R_k R_{k-1} \cdots R_1 B^+\{a_n\}_{n=N}^\infty$$

or

$$\{b_n\}_{n=N}^\infty = R_k R_{k-1} \cdots R_1 B^-\{a_n\}_{n=N}^\infty.$$

Then, if $\lim_{n\to\infty} b_n > 1$, $\sum_{n=N}^\infty a_n$ *converges; if* $\lim_{n\to\infty} b_n < 1$, $\sum_{n=N}^\infty a_n$ *diverges.*

More generally, if $\liminf_{n\to\infty} b_n > 1$ *(resp.* $\limsup_{n\to\infty} b_n < 1$*),* $\sum_n^\infty a_n$ *converges (resp. diverges).*

2.5 Tests of Abel

We turn now to a collection of tests that follow quickly from an algebraic manipulation due to Abel. Abelian summation has applications beyond infinite series. It is often applicable to the finite summations of combinatorics, for example. The name "summation by parts" comes from an analogy with the "integration by parts" technique of integral calculus. In so-called difference calculus, the analogue of integration is the partial sum, and the analogue of differentiation is the first difference. Viewed in that perspective, the following result has almost exactly the same form as integration by parts. It tends to be useful in the study of infinite series in much the same way that integration by parts is useful in calculus—whenever series that are understood are combined by term-by-term multiplication.

Theorem 2.20 (Abel, Summation by Parts) *Let $\{a_n\}_{n=0}^{\infty}, \{b_n\}_{n=0}^{\infty}$ be arbitrary sequences, and write*

$$A_n = a_0 + a_1 + a_2 + \cdots + a_n$$

for the partial sums of $\sum_{n=0}^{\infty} a_n$, where we include the degenerate case $A_{-1} = 0$.

Then, for each $0 \leq m \leq n$,

$$\sum_{k=m}^{n} a_k b_k = \left[\sum_{k=m}^{n} A_k(b_k - b_{k+1})\right] - A_{m-1}b_m + A_n b_{n+1}.$$

Proof. First write

$$a_k b_k = (A_k - A_{k-1})b_k = A_k b_k - A_k b_{k+1} + (A_k b_{k+1} - A_{k-1}b_k).$$

The stated result follows by summing this equality from m to n and recognizing that the final two terms telescope.

Theorem 2.21 *Let $\{a_n\}_{n=0}^{\infty}, \{b_n\}_{n=0}^{\infty}$ be arbitrary sequences, and write*

$$A_n = a_0 + a_1 + a_2 + \cdots a_n$$

for the partial sums of $\sum_{n=0}^{\infty} a_n$, where we include the degenerate case $A_{-1} = 0$.

Then, if $\sum_{n=0}^{\infty} A_n(b_n - b_{n+1})$ converges and $\lim_{n\to\infty} A_n b_{n+1}$ exists, then the series $\sum_{n=0}^{\infty} a_n b_n$ converges.

Proof. By the preceding theorem, the partial sums s_k for the series $\sum_{n=0}^{\infty} a_n b_n$ are given by

$$\begin{aligned} s_k = \sum_{n=0}^{k} a_n b_n &= \left[\sum_{n=0}^{k} A_n(b_n - b_{n+1})\right] - A_{-1}b_0 + A_k b_{k+1} \\ &= \left[\sum_{n=0}^{k} A_n(b_n - b_{n+1})\right] + A_k b_{k+1}. \end{aligned}$$

Then the sequence of partial sums will certainly converge if each of the two summands on the right-hand side converges independently. But this is precisely the statement of the theorem.

There are a number of corollaries to this result, and it is no longer clear to whom each should rightly be attributed. Here we give some names commonly attached to these tests. In addition to Abel, we identify the mathematicians Richard Dedekind (1831–1916), Lejeune Dirichlet (1805–1859), and Paul David du Bois-Reymond (1831–1889).

Corollary 2.21.1 (Abel) *Let $\sum_{n=0}^{\infty} a_n$ be a convergent series, and let $\{b_n\}_{n=0}^{\infty}$ be a bounded monotone sequence. Then $\sum_{n=0}^{\infty} a_n b_n$ converges.*

Proof. We will check the two hypotheses in order to apply Theorem 2.21. The A_n in that theorem are exactly the partial sums of the convergent series $\sum_{n=0}^{\infty} a_n$, hence we can write

$$A = \lim_{n\to\infty} A_n.$$

Also, since $\{b_n\}_{n=0}^{\infty}$ is monotone and bounded, it must be a convergent sequence, so we write

$$B = \lim_{n\to\infty} b_n$$

and notice that, by telescoping,

$$\sum_{n=0}^{\infty}(b_n - b_{n+1}) = b_0 - B$$

is a convergent series.

Then, because the above series has terms all of the same sign, and because A_n is convergent (hence bounded), we can say that $\sum_{n=0}^{\infty} A_n(b_n - b_{n+1})$ is also convergent by the Limit Comparison Test Strengthened. Finally,

$$\lim_{n\to\infty} A_n b_{n+1} = AB,$$

so this limit also exists. We can now apply Theorem 2.21.

Example 2.22 (Compare Example 2.2) *Suppose that $\sum_{n=1}^{\infty} a_n$ is a convergent series of positive terms, and let $r_n = \sum_{k=n}^{\infty} a_k$ be the nth tail for each n.*

$$\sum_{n=1}^{\infty} a_n r_n$$

converges.

Solution. To use Corollary 2.21.1, we only have to notice that the sequence $\{r_k\}$ is monotone decreasing, and convergent to 0 (hence bounded). Since $\sum_{n=1}^{\infty} a_n$ converges, the corollary applies and we are done.

Example 2.23 *Suppose that $\sum_{n=1}^{\infty} a_n$ is a convergent series of positive terms, and let $s_n = \sum_{k=1}^{n} a_k$ be the nth partial sum for each n.*

$$\sum_{n=1}^{\infty} a_n s_n$$

converges, and so also does

$$\sum_{n=1}^{\infty} \frac{a_n}{s_n}.$$

Solution. To use Corollary 2.21.1, we only have to notice that the sequence $\{s_k\}$ is monotone increasing, and convergent to $\sum_{n=1}^{\infty} a_n$ (hence bounded); also the series $\{1/s_k\}$ is monotone decreasing and bounded. Since $\sum_{n=1}^{\infty} a_n$ converges, the corollary applies and we are done.

Corollary 2.23.1 (Dirichlet) *Let $\sum_{n=0}^{\infty} a_n$ be a series with bounded partial sums, and let $\{b_n\}_{n=0}^{\infty}$ be a monotone sequence with $\lim_{n\to\infty} b_n = 0$. Then $\sum_{n=0}^{\infty} a_n b_n$ converges.*

Proof. We can show that $\sum_{n=0}^{\infty} A_n(b_n - b_{n+1})$ converges by precisely the same argument as the last corollary. Now, because the A_n are bounded, there exist constants E_1, E_2 such that

$$E_1 b_{n+1} \le A_n b_{n+1} \le E_2 b_{n+1},$$

making it clear that $A_n b_{n+1} \to 0$ also.

Corollary 2.23.2 (Dedekind, Bois-Reymond) *Let $\sum_{n=0}^{\infty} a_n$ be a convergent series, and let $\{b_n\}_{n=0}^{\infty}$ be such that $\sum_{n=0}^{\infty}(b_n - b_{n+1})$ converges absolutely. Then $\sum_{n=0}^{\infty} a_n b_n$ converges.*

Proof. Follow the proof of Abel's Corollary, except that now we can show that

$$\sum_{n=0}^{\infty} A_n(b_n - b_{n+1})$$

is *absolutely* convergent by the Limit Comparison Test Strengthened.

Example 2.24 *Investigate the convergence of*

$$\sum_{n=2}^{\infty} \frac{\tan^{-1}(n)}{n(\log n)^2}.$$

Solution. We first recognize that the denominator of the expression for a general term resembles the Abel-Dini scale. This suggests writing out terms as a product of simpler expressions

$$\sum_{n=2}^{\infty} \frac{1}{n(\log n)^2} \tan^{-1}(n).$$

Since

$$\sum_{n=2}^{\infty} \frac{1}{n(\log n)^2}$$

converges and

$$\tan^{-1}(n)$$

is increasing in n with $\tan^{-1}(n) \leq \pi/2$, we can apply Corollary 2.21.1. For another example of this type of argument in action, see Gem 102.

Appendix to Chapter 2: Proofs of Bertrand's Tests

In this appendix, we give proofs for Theorems 2.17, 2.18, and 2.19. The proofs here are considerably longer and more technical than the other material in this book, so the reader should feel free to just skip this section and accept the truth of Bertrand's tests. This proof has been included for completeness, due to the lack of any good reference in the literature.

We begin with some notation and computation regarding iterated logarithms, some of them familiar.

Notation 2.25 *Let $\log_{(k)} x$ be the iterated natural logarithm as defined earlier. We also write $\Lambda_{m,n}(x)$ for $\prod_{i=m}^{n} \log_{(i)} x$ and $\Omega_{m,n}(x)$ for*

$$\sum_{k=m}^{n} \frac{1}{\Lambda_{0,k}(x)}$$

for nonnegative integers $m \leq n$.

Notation 2.26 *It is understood that these functions (and any functions defined in terms of these) are only defined for x sufficiently large. Any relation in which such functions appear is to be interpreted only "eventually" with respect to the appropriate variable. In that spirit, lower bounds will typically be omitted from infinite integrals and summations, it being understood that the choice does not affect convergence.*

Notation 2.27 *An "empty" summation in which the lower limit exceeds the upper limit will be interpreted as having the value 0; an empty product has the value 1. In particular we extend $\Lambda_{m,n}(x)$ and $\Omega_{m,n}(x)$ to include such cases.*

Proposition 2.28

$$D_x \log_{(k)}(x) = \frac{1}{\Lambda_{0,k-1}(x)}.$$

Proof. By induction on k. The cases $k = 0, 1$ are the trivial $D_x x = 1$ and $D_x(\log x) = 1/x$. Assume that the statement holds for some $k \geq 1$. Then by the Chain Rule,

$$\begin{aligned}(\log_{(k+1)})'(x) = (\log \log_{(k)})'(x) &= \frac{1}{\log_{(k)}(x)}(\log_{(k)})'(x) \\ &= \frac{1}{\log_{(k)}(x)\Lambda_{0,k-1}(x)} = \frac{1}{\Lambda_{0,k}(x)},\end{aligned}$$

completing the induction.

Proposition 2.29

$$D_x \Lambda_{m,n}(x) = \Lambda_{m,n}(x)\Omega_{m,n}(x).$$

Proof. We apply the Product Rule and the preceding proposition.

$$\begin{aligned}
D_x \Lambda_{m,n}(x) &= D_x \prod_{k=m}^{n} \log_{(k)}(x) \\
&= \sum_{j=m}^{n} \left[\left(\prod_{k=m}^{j-1} \log_{(k)}(x) \right) (D_x \log_j(x)) \left(\prod_{k=j+1}^{n} \log_{(k)}(x) \right) \right] \\
&= \sum_{j=m}^{n} \left[\left(\prod_{k=m}^{j-1} \log_{(k)}(x) \right) \left(\frac{1}{\Lambda_{0,j-1}(x)} \right) \left(\prod_{k=j+1}^{n} \log_{(k)}(x) \right) \right] \\
&= \sum_{j=m}^{n} \left[\left(\prod_{k=m}^{j-1} \log_{(k)}(x) \right) \left(\frac{\log_j(x)}{\Lambda_{0,j}(x)} \right) \left(\prod_{k=j+1}^{n} \log_{(k)}(x) \right) \right] \\
&= \sum_{j=m}^{n} \left(\prod_{k=m}^{n} \log_{(k)}(x) \frac{1}{\Lambda_{0,j}(x)} \right) = \Lambda_{m,n}(x)\Omega_{m,n}(x).
\end{aligned}$$

Proposition 2.30

$$D_x[\Lambda_{m,n}(x)(\log_{(n)} x)^\alpha] = \frac{\left[\left(\sum_{i=m}^{n-1} \Lambda_{i+1,n-1}(x)(\log_{(n)}(x))^\alpha \right) + \alpha(\log_{(n)}(x))^{\alpha-1} \right]}{\Lambda_{0,m-1}(x)}.$$

Proof. We apply the Product Rule, the Chain Rule, and the preceding propositions.

$$\begin{aligned}
D_x[\Lambda_{m,n}(x)(\log_{(n)} x)^\alpha] &= (\log_{(n)} x)^\alpha D_x \Lambda_{m,n}(x) + \Lambda_{m,n-1}(x) D_x (\log_{(n)} x)^\alpha \\
&= (\log_{(n)} x)^\alpha [\Lambda_{m,n-1}(x)\Omega_{m,n-1}(x)] \\
&\quad + \Lambda_{m,n-1}(x)(\log_{(n)} x)^{\alpha-1} \frac{\alpha}{\Lambda_{0,n-1}(x)} \\
&= \Lambda_{m,n-1}(x)(\log_{(n)} x)^\alpha \left(\Omega_{m,n-1}(x) + \frac{\alpha}{\Lambda_{0,n}(x)} \right) \\
&= \frac{\left[\left(\sum_{i=m}^{n-1} \Lambda_{i+1,n-1}(x)(\log_{(n)}(x))^\alpha \right) + \alpha(\log_{(n)}(x))^{\alpha-1} \right]}{\Lambda_{0,m-1}(x)}.
\end{aligned}$$

Notation 2.31 *Let $f(x)$, $g(x)$ be defined for all x sufficiently large, with $g(x)$ eventually positive. Then we write $f(x) = o(g(x))$ to mean*

$$\lim_{x \to \infty} \frac{f(x)}{g(x)} = 0.$$

Notation 2.32 *Let $f(x)$, $g(x)$ be defined for x sufficiently large, with $g(x)$ eventually positive. Then we write $f(x) = O(g(x))$ to mean that, for some constant C and for all x*

sufficiently large,

$$|f(x)| \leq Cg(x).$$

Proposition 2.33

$$(\log_{(m)} x)^\alpha = o((\log_{(n)} x)^\beta)$$

for any integers $m > n \geq 0$ and any exponents α, β provided $\beta > 0$.

Proof. If $\alpha \leq 0$, then this is trivial, so we suppose $\alpha > 0$. Using the substitution $u = \log_{(n)} x$ and L'Hôpital's rule,

$$\begin{aligned}
\lim_{x\to\infty} \frac{(\log_{(m)} x)^\alpha}{(\log_{(n)} x)^\beta} &= \lim_{u\to\infty} \frac{(\log_{(m-n)} u)^\alpha}{u^\beta} \\
&= \lim_{x\to\infty} \left(\frac{\log_{(m-n)} u}{u^{\beta/\alpha}} \right)^\alpha \\
&= \lim_{x\to\infty} \left(\frac{\frac{1}{\Lambda_{0,m-n-1}(u)}}{(\beta/\alpha) u^{\beta/\alpha - 1}} \right)^\alpha \\
&= \lim_{x\to\infty} \left(\frac{1}{(\beta/\alpha)\Lambda_{1,m-n-1}(u) u^{\beta/\alpha}} \right)^\alpha \\
&= 0
\end{aligned}$$

as desired.

Proposition 2.34 *Let k be a nonnegative integer and let α be real. Then*

$$\frac{\Lambda_{0,k-1}(n+1)(\log_{(k)}(n+1))^\alpha}{\Lambda_{0,k-1}(n)(\log_{(k)} n)^\alpha} = 1 + \Omega_{0,k-1}(n) + \frac{\alpha}{\Lambda_{0,k}(n)} + o\left(\frac{1}{\Lambda_{0,k}(n)}\right).$$

Proof. In the case where $k = 0$, we use the Binomial Theorem to obtain

$$\frac{(n+1)^\alpha}{n^\alpha} = \left(1 + \frac{1}{n}\right)^\alpha = 1 + \frac{\alpha}{n} + O\left(\frac{1}{n^2}\right) = 1 + \frac{\alpha}{n} + o\left(\frac{1}{n}\right).$$

Assume henceforth that $k > 0$. By the Mean Value Theorem and Proposition 2.30,

$$\begin{aligned}
\Lambda_{0,k-1}(n+1)(\log_{(k)}(n+1))^\alpha &= \Lambda_{0,k-1}(n)(\log_{(k)} n)^\alpha \\
&\quad + \sum_{i=0}^{k-1} \Lambda_{i+1,k-1}(n')(\log_{(k)} n')^\alpha + \alpha(\log_{(k)} n')^{\alpha-1}
\end{aligned}$$

for some $n < n' < n + 1$. Another application of the Mean Value Theorem gives

$$\sum_{i=0}^{k-1} \Lambda_{i+1,k-1}(n')(\log_{(k)} n')^{\alpha} = \sum_{i=0}^{k-1} \Lambda_{i+1,k-1}(n)(\log_{(k)} n)^{\alpha}$$

$$+ \left(\frac{(n'-n)}{\Lambda_{0,i}(n_i)} \left[\left(\sum_{j=i+1}^{k-1} \Lambda_{j+1,k-1}(n_i)(\log_{(k)} n_i)^{\alpha} \right) + \alpha(\log_{(k)} n_i)^{\alpha} \right] \right)$$

for each $i = 0, 1, \ldots, k-1$, where $n < n_i < n'$. But then using Proposition 2.33,

$$\Lambda_{i+1,k-1}(n')(\log_{(k)} n')^{\alpha} = \Lambda_{i+1,k-1}(n)(\log_{(k)} n)^{\alpha} + o((\log_{(k)} n)^{\alpha-1}).$$

Similarly, for some $n < n'' < n'$,

$$\begin{aligned}
(\log_{(k)} n')^{\alpha-1} &= (\log_{(k)} n)^{\alpha-1} + (n'-n)\frac{(\alpha-1)(\log_{(k)}(n''))^{\alpha-1}}{\Lambda_{0,k-1}n''} \\
&= (\log_{(k)} n)^{\alpha-1} + O(n^{-1})o(n^{-1/2}) \\
&= (\log_{(k)} n)^{\alpha-1} + o(n^{1/2}) \\
&= (\log_{(k)} n)^{\alpha-1} + o((\log_{(k)} n)^{\alpha-1}).
\end{aligned}$$

Combining all these equations,

$$\begin{aligned}
\Lambda_{0,k-1}(n+1)(\log_{(k)}(n+1))^{\alpha} = {} & \Lambda_{0,k-1}(n)(\log_{(k)} n)^{\alpha} \\
& + \sum_{i=0}^{k-1} \Lambda_{i+1,k-1}(n')(\log_{(k)} n')^{\alpha} \\
& + \alpha(\log_{(k)} n)^{\alpha-1} + o((\log_{(k)} n)^{\alpha-1})
\end{aligned}$$

and dividing through by $\Lambda_{0,k-1}(n)(\log_{(k)} n)^{\alpha}$ gives exactly what is desired.

At last we are in position to prove Bertrand's sequence of tests.

Theorem 2.35 (Restatement of Theorem 2.17) *Let $\sum_n^{\infty} a_n$ be a series of positive real numbers. Let k be a nonnegative integer, and define a sequence $\{b_n\}$ implicitly by*

$$\frac{a_n}{a_{n+1}} = 1 + \Omega_{0,k-1}(n) + \frac{b_n}{\Lambda_{0,k}(n)}.$$

Then if $\liminf_{n\to\infty} b_n > 1$, *$\sum_n^{\infty} a_n$ converges; if* $\limsup_{n\to\infty} b_n < 1$, *$\sum_n^{\infty} a_n$ diverges.*

Proof. We will consider only the convergence half of the theorem; the proof of the divergence half is analogous.

Suppose then that $\liminf_{n\to\infty} b_n > 1$. Then there exists a real number α such that $1 < \alpha < b_n$ eventually. By Theorem 2.16, the series $\sum_n^{\infty} c_n$ with

$$c_n = \frac{1}{\Lambda_{0,k-1}(n)(\log_{(k)} n)^{\alpha}}$$

converges. Since $\alpha < b_n$ eventually, by Proposition 2.34 we eventually have

$$\begin{aligned}\frac{c_{n+1}}{c_n} &= \left(1 + \Omega_{0,k-1}(n) + \frac{\alpha + o(1)}{\Lambda_{0,k}(n)}\right)^{-1} \\ &> \left(1 + \Omega_{0,k-1}(n) + \frac{b_n}{\Lambda_{0,k}(n)}\right)^{-1} \\ &= \frac{a_{n+1}}{a_n}.\end{aligned}$$

So $\sum_n^\infty a_n$ converges by the Ratio Comparison Test.

In a very similar fashion, we can prove the second form.

Proposition 2.36 *Let k be a nonnegative integer and let α be real. Then*

$$\frac{\Lambda_{0,k-1}(n)(\log_{(k)} n)^\alpha}{\Lambda_{0,k-1}(n+1)(\log_{(k)}(n+1))^\alpha} = 1 - \Omega_{0,k-1}(n) - \frac{\alpha}{\Lambda_{0,k}(n)} + o\left(\frac{1}{\Lambda_{0,k}(n)}\right).$$

Proof. The proof follows the same lines as Proposition 2.34, appealing to the Mean Value Theorem repeatedly. We will not repeat this work here.

Theorem 2.37 (Restatement of Theorem 2.18) *Let $\sum_n^\infty a_n$ be a series of positive real numbers. Let k be a nonnegative integer, and define a sequence $\{b_n\}$ implicitly by*

$$\frac{a_{n+1}}{a_n} = 1 - \Omega_{0,k-1}(n) - \frac{b_n}{\Lambda_{0,k}(n)}.$$

Then if $\liminf_{n\to\infty} b_n > 1$, *$\sum_n^\infty a_n$ converges; if* $\limsup_{n\to\infty} b_n < 1$, *$\sum_n^\infty a_n$ diverges.*

Proof. We will consider only the convergence half of the theorem; the proof of the divergence half is analogous.

Suppose then that $\liminf_{n\to\infty} b_n > 1$. Then there exists a real number α such that $1 < \alpha < b_n$ eventually. By Theorem 2.16, the series $\sum_n^\infty c_n$ with

$$c_n = \frac{1}{\Lambda_{0,k-1}(n)(\log_{(k)} n)^\alpha}$$

converges. Since $\alpha < b_n$ eventually, by Proposition 2.36 we eventually have

$$\frac{c_{n+1}}{c_n} = \left(1 - \Omega_{0,k-1}(n) - \frac{\alpha + o(1)}{\Lambda_{0,k}(n)}\right) > \left(1 - \Omega_{0,k-1}(n) - \frac{b_n}{\Lambda_{0,k}(n)}\right) = \frac{a_{n+1}}{a_n}.$$

So $\sum_n^\infty a_n$ converges by the Ratio Comparison Test.

Theorem 2.38 (Restatement of Theorem 2.19) *Define operators B^+, B^- which act on sequences by*

$$B^+\{b_n\}_{n=N}^{\infty} = \left\{n(\frac{b_n}{b_{n+1}} - 1)\right\}_{n=N}^{\infty}$$

$$B^-\{b_n\}_{n=N}^{\infty} = \left\{n(1 - \frac{b_{n+1}}{b_n})\right\}_{n=N}^{\infty}.$$

For each positive integer k, let R_k be the operator defined by

$$R_k\{b_n\}_n^{\infty} = \{\log_{(k)}(n)(b_n - 1)\}_n^{\infty}.$$

Let $\sum_n^{\infty} a_n$ be a series of positive real numbers. Let k be a nonnegative integer, and define a sequence $\{b_n\}_n^{\infty}$ by either

$$\{b_n\}_{n=N}^{\infty} = R_k R_{k-1} \cdots R_1 B^+\{a_n\}_{n=N}^{\infty}$$

or

$$\{b_n\}_{n=N}^{\infty} = R_k R_{k-1} \cdots R_1 B^-\{a_n\}_{n=N}^{\infty}.$$

Then, if $\lim_{n\to\infty} b_n > 1$, $\sum_{n=N}^{\infty} a_n$ converges; if $\lim_{n\to\infty} b_n < 1$, $\sum_{n=N}^{\infty} a_n$ diverges.

More generally, if $\liminf_{n\to\infty} b_n > 1$(resp. $\limsup_{n\to\infty} b_n < 1$), $\sum_{n=N}^{\infty} a_n$ converges (resp. diverges).

Proof. First, observe that

$$\frac{a_n}{a_{n+1}} = 1 + \Omega_{0,k-1}(n) + \frac{R_k R_{k-1} \cdots R_1 B^+\{a_n\}_{n=N}^{\infty}}{\Lambda_{0,k}(n)}$$

and that

$$\frac{a_{n+1}}{a_n} = 1 + \Omega_{0,k-1}(n) + \frac{R_k R_{k-1} \cdots R_1 B^-\{a_n\}_{n=N}^{\infty}}{\Lambda_{0,k}(n)}.$$

These formal identities are totally straightforward to prove, once guessed, using induction on k. We will prove only the first one here, leaving the second to the reader.

The case $k = 0$ is merely the statement

$$\frac{a_n}{a_{n+1}} = 1 + \frac{n(a_n/a_{n+1} - 1)}{n} = 1 + \frac{B^+(\{a_n\})}{n}.$$

Now suppose that the identity holds for some $k \geq 0$. Then

$$\begin{aligned}
1 + \Omega_{0,k}(n) &+ \frac{R_{k+1} R_k \cdots R_1 B^+\{a_n\}_{n=N}^{\infty}}{\Lambda_{0,k+1}(n)} \\
&= 1 + \Omega_{0,k-1}(n) + \frac{1}{\Lambda_{0,k}(n)} + \frac{R_{k+1} R_k \cdots R_1 B^+\{a_n\}_{n=N}^{\infty}}{\Lambda_{0,k+1}(n)} \\
&= 1 + \Omega_{0,k-1}(n) + \frac{1}{\Lambda_{0,k}(n)} + \frac{\log_{(k+1)}(n)[R_k \cdots R_1 B^+\{a_n\}_{n=N}^{\infty} - 1]}{\Lambda_{0,k+1}(n)}
\end{aligned}$$

$$= 1 + \Omega_{0,k-1}(n) + \frac{1}{\Lambda_{0,k}(n)} + \frac{R_k \cdots R_1 B^+\{a_n\}_{n=N}^{\infty} - 1}{\Lambda_{0,k}(n)}$$

$$= 1 + \Omega_{0,k-1}(n) + \frac{R_k \cdots R_1 B^+\{a_n\}_{n=N}^{\infty}}{\Lambda_{0,k}(n)}$$

$$= \frac{a_n}{a_{n+1}},$$

completing the induction.

In the light of this observation, the claim is just a rephrasing of the preceding two theorems.

3

The Harmonic Series and Related Results

This chapter is dedicated to the harmonic series, which is the series of reciprocals of the positive integers.

$$\sum_{n=1}^{\infty} \frac{1}{n} = 1 + \frac{1}{2} + \frac{1}{3} + \frac{1}{4} + \cdots$$

The harmonic series has been the object of much interest and many interesting results, most likely because it is in many senses the series with the simplest possible terms whose terms converge to zero. This is interesting because, although the harmonic series has terms tending to zero, the series diverges. Before proceeding to some more curious results, we prove the divergence of this series in several ways. It is worth taking our time about this to ensure that the reader really believes and sees clearly why this series diverges. Calculus students first encountering infinite series tend to pass through two stages of faulty reasoning about series. The first stage is the belief that an infinite series cannot possibly have a finite sum, as discussed in Chapter 1 in the context of Zeno, who like the rest of his generation was still in this stage. The second stage is an overcompensation to the first—students in this stage believe not only the fact that the terms of a convergent series must approach zero but also its converse. However, we have already pointed out that the Divergence Test has a false converse. We present the harmonic series as the classic counterexample, though we have already seen other counterexamples. It is only after passing through both of these faulty stages that real intuitive reasoning about series can take place.

Though every book mentioning infinite series at all almost certainly mentions the harmonic series, very seldom is the origin of its name discussed. Naively, we would expect that the word "harmonic" has its roots in "harmony," and in fact this is correct. Consider a very simple musical instrument for the purposes of discussion, the 1-string harp. (That is, consider a vibrating string with both ends fixed.) When this string is plucked, we often visualize its vibration as a single wave. This is the simplest kind of vibration and is called the *first harmonic*. However, it is also possible for the string to vibrate in two pieces (the *second harmonic*). In this case the wavelength, which is directly related to the pitch of the sound, is $1/2$ the wavelength of the first harmonic. Indeed, for each integer n the string can vibrate so that n arcs are formed, and the wavelength will be $1/n$ times the wavelength for the first harmonic. In this way the series of reciprocals relates to harmonics and harmonies. In fact it is possible for the same string to vibrate in combinations of these modes, in which (say) the third harmonic overlays the second harmonic. Combinations of this sort lead to multiple tones and hence to *harmonies*. In a sense looking at harmonics is the *right* way to look at harmony, because, if we change the length of the string, the tones associated to each harmonic change, but the tonic relationship of (say) the third and second harmonics remains analogous. We should also mention that the harmonic series is related to the harmonic mean. Just as each term of an arithmetic series is the arithmetic mean of the terms immediately preceding and following, each term in the harmonic series is the harmonic mean of the surrounding terms. The interested reader can read more about the origin of the term in an article entitled "What's Harmonic about the Harmonic Series?" by David E. Kullman in *CMJ*, May 2001, Vol. 32, No. 3, pp. 201–3. This article is reproduced in Appendix B of this book.

3.1 Divergence Proofs

As already mentioned, there are good intuitive reasons to suspect that the harmonic series converges. How could it diverge, the naive observer may well wonder, if the terms you are adding are getting smaller and smaller? To settle this question in the reader's mind once and for all, we include not one, not two, but twelve proofs that the harmonic series diverges. (One of the twelve proofs is later in the chapter.) The reader is invited to notice the range of techniques used by different mathematicians to show this fundamental result, and furthermore to notice the relative elegance of certain proofs over others.

Theorem 3.1

$$\sum_{n=1}^{\infty} \frac{1}{n} = \infty,$$

that is, the harmonic series diverges.

Proof 1. This is the standard proof exhibited by many calculus books, often called the Cauchy proof (because it is a special case of the proof of Cauchy's Condensation Test). However, this proof of the divergence of the harmonic series can actually be traced back to the Frenchman Nicole Oresme. The proof, like most proofs that we exhibit, demonstrates that the partial sums increase without bound. We denote the nth partial sum by S_n. Now we notice that

$$S_2 = 1 + \frac{1}{2} > \frac{1}{2} + \frac{1}{2} = 1$$
$$S_4 = S_2 + \frac{1}{3} + \frac{1}{4} > 1 + \frac{1}{4} + \frac{1}{4} = 1 + \frac{1}{2} = \frac{3}{2}$$
$$S_8 = S_4 + \frac{1}{5} + \frac{1}{6} + \frac{1}{7} + \frac{1}{8} > \frac{3}{2} + \frac{1}{8} + \frac{1}{8} + \frac{1}{8} + \frac{1}{8} = \frac{3}{2} + \frac{1}{2} = 2.$$

Proceeding in the same way, we find that $S_{2^n} > (n+1)/2$. Since the right-hand side of this inequality clearly increases to infinity, the harmonic series cannot converge.

A more visually pleasing way to see this is as follows.

$$\sum_{n=1}^{\infty} a_n = 1 + \frac{1}{2} + \frac{1}{3} + \frac{1}{4} + \frac{1}{5} + \frac{1}{6} + \frac{1}{7} + \frac{1}{8} + \frac{1}{9} + \frac{1}{10} + \cdots + \frac{1}{16} + \frac{1}{17} + \cdots$$
$$\sum_{n=1}^{\infty} b_n = 1 + \frac{1}{2} + \frac{1}{4} + \frac{1}{4} + \frac{1}{8} + \frac{1}{8} + \frac{1}{8} + \frac{1}{8} + \frac{1}{16} + \frac{1}{16} + \cdots + \frac{1}{16} + \frac{1}{32} + \cdots +$$
$$= 1 + \frac{1}{2} + \frac{1}{2} + \frac{1}{2} + \frac{1}{2} + \frac{1}{2} + \cdots,$$

to which we can apply the standard comparison test.

As remarked earlier, this is identical in substance to Cauchy's Condensation Test, and one can just apply that theorem instead of working through this proof.

Proof 2. We appeal directly to the Integral Test of Chapter 1. Observe that $\int_1^{\infty} \frac{1}{x}\,dx$ is a divergent improper integral whose integrand is positive, continuous, and monotone decreasing, so $\sum_{n=1}^{\infty} 1/n$ is a divergent series. (Note: This was the argument used in Chapter 1 to prove that the harmonic series diverged as a p-series with $p = 1$.)

Proof 3 [CMJ, 28:3, pp. 209–210]. Assume for the sake of contradiction that the harmonic series converges to a sum S. Consider two auxiliary series, the "odd harmonic" and "even harmonic" series, which have terms given (respectively) by

$$1 + \frac{1}{3} + \frac{1}{5} + \frac{1}{7} + \cdots = \sum_{n=1}^{\infty} \frac{1}{2n-1}$$
$$\frac{1}{2} + \frac{1}{4} + \frac{1}{6} + \frac{1}{8} + \cdots = \sum_{n=1}^{\infty} \frac{1}{2n}.$$

Since $1/n \geq 1/(2n-1) > 1/(2n)$, we can use the original Comparison Test to see that, if the harmonic series converges to S these auxiliary series converge also to sums S_o, S_e satisfying

$$S \geq S_o > S_e.$$

Each term in the harmonic series appears in exactly one of the auxiliary series, so we can see that

$$S = S_o + S_e,$$

taking into account that we can freely scramble and group the terms in a convergent series of positive terms without affecting its convergence or its sum. On the other hand, each term in the even harmonic series is half the corresponding term in the harmonic series so that

$$S_e = \frac{1}{2}S.$$

Putting these together gives

$$S_o = S - S_e = \frac{1}{2}S = S_e,$$

a contradiction.

Proof 4 [CMJ, 30:1, p. 34]. First we make the computation

$$\frac{1}{n} + \frac{1}{n+1} = \frac{2n+1}{n(n+1)} = \frac{2n}{n(n+1)} + \frac{1}{n(n+1)} = \frac{1}{(n+1)/2} + \frac{1}{n(n+1)}.$$

Now as before we assume for the sake of contradiction that the harmonic series converges to a sum S. Then we can write:

$$\begin{aligned} S &= 1 + \frac{1}{2} + \frac{1}{3} + \frac{1}{4} + \frac{1}{5} + \frac{1}{6} + \frac{1}{7} + \frac{1}{8} + \cdots \\ &= \left(1 + \frac{1}{2}\right) + \left(\frac{1}{3} + \frac{1}{4}\right) + \left(\frac{1}{5} + \frac{1}{6}\right) + \left(\frac{1}{7} + \frac{1}{8}\right) + \cdots \\ &= \left(1 + \frac{1}{2}\right) + \left(\frac{1}{2} + \frac{1}{12}\right) + \left(\frac{1}{3} + \frac{1}{30}\right) + \left(\frac{1}{4} + \frac{1}{56}\right) + \cdots \\ &= S + \left(\frac{1}{2} + \frac{1}{12} + \frac{1}{30} + \frac{1}{56} + \cdots\right) \\ &> S \end{aligned}$$

for our contradiction.

Proof 5. Once more we assume that the harmonic series converges with sum S. Remembering that grouping does not change the sum of a convergent series, we group the terms into blocks of n terms (where we can freely choose any $n \geq 2$). Choosing for example $n = 4$, we have

$$\begin{aligned} S &= \left(1 + \frac{1}{2} + \frac{1}{3} + \frac{1}{4}\right) + \left(\frac{1}{5} + \frac{1}{6} + \frac{1}{7} + \frac{1}{8}\right) + \left(\frac{1}{9} + \frac{1}{10} + \frac{1}{11} + \frac{1}{12}\right) + \cdots \\ &> \left(\frac{1}{4} + \frac{1}{4} + \frac{1}{4} + \frac{1}{4}\right) + \left(\frac{1}{8} + \frac{1}{8} + \frac{1}{8} + \frac{1}{8}\right) + \left(\frac{1}{12} + \frac{1}{12} + \frac{1}{12} + \frac{1}{12}\right) + \cdots \end{aligned}$$

$$
\begin{aligned}
&= 1 + \frac{1}{2} + \frac{1}{3} + \cdots + \\
&= S
\end{aligned}
$$

for another contradiction.

Proof 6. This proof is known as Mengoli's Divergence proof and somewhat resembles the $n = 3$ case of the previous proof. Like Proof 4, the proof depends on an algebraic manipulation.

$$
\frac{1}{p-1} + \frac{1}{p} + \frac{1}{p+1} = \frac{2p}{p^2-1} + \frac{1}{p} > \frac{2p}{p^2} + \frac{1}{p} = \frac{3}{p}.
$$

Once again we assume for the sake of contradiction that the harmonic series converges to S, and rewrite as follows:

$$
\begin{aligned}
S &= 1 + \left(\frac{1}{2} + \frac{1}{3} + \frac{1}{4}\right) + \left(\frac{1}{5} + \frac{1}{6} + \frac{1}{7}\right) + \left(\frac{1}{8} + \frac{1}{9} + \frac{1}{10}\right) + \cdots \\
&> 1 + 3\left(\frac{1}{3}\right) + 3\left(\frac{1}{6}\right) + 3\left(\frac{1}{9}\right) + \cdots \\
&= 1 + 1 + \frac{1}{2} + \frac{1}{3} + \cdots \\
&= 1 + S.
\end{aligned}
$$

Once more a contradiction arises from assuming that S is a finite number.

Proof 7. This proof originates with Jakob Bernoulli in 1689. We first observe that for any integer $p > 1$ we have

$$
\sum_{n=p+1}^{p^2} \frac{1}{n} > \sum_{n=p+1}^{p^2} \frac{1}{p^2} = \frac{p^2 - p}{p^2} = 1 - \frac{1}{p},
$$

so that in particular

$$
\sum_{n=p}^{p^2} \frac{1}{n} > 1.
$$

The trick now is to group the terms into blocks so that the above observation applies to each block. To that end we define a sequence of numbers by $a_1 = 2$, $a_{n+1} = a_n^2 + 1$. Then, by grouping, we have

$$
\sum_{n=1}^{\infty} \frac{1}{n} = 1 + \sum_{n=a_1}^{a_1^2} \frac{1}{n} + \sum_{n=a_2}^{a_2^2} \frac{1}{n} + \sum_{n=a_3}^{a_3^2} \frac{1}{n} + \cdots > 1 + 1 + 1 + 1 + \cdots,
$$

so that the harmonic series diverges.

Proof 8 [CMJ, 18:1, pp. 18–23]. This proof is also due to a Bernoulli in 1689, this time Johann. Bernoulli made the observation that

$$\sum_{n=1}^{\infty} \frac{1}{n(n+1)}$$

is a telescoping series that converges to $C = 1$ (see Example 1.42). We write

$$\begin{array}{llll}
C = \frac{1}{2} + \frac{1}{6} + \frac{1}{12} + \frac{1}{20} + \frac{1}{30} + \cdots = & & 1 = 1 \\
D = \phantom{\frac{1}{2} +} \frac{1}{6} + \frac{1}{12} + \frac{1}{20} + \frac{1}{30} + \cdots = & C - \frac{1}{2} & = \frac{1}{2} \\
E = \phantom{\frac{1}{2} + \frac{1}{6} +} \frac{1}{12} + \frac{1}{20} + \frac{1}{30} + \cdots = & D - \frac{1}{6} & = \frac{1}{3}. \\
F = \phantom{\frac{1}{2} + \frac{1}{6} + \frac{1}{12} +} \frac{1}{20} + \frac{1}{30} + \cdots = & E - \frac{1}{12} & = \frac{1}{4} \\
G = \phantom{\frac{1}{2} + \frac{1}{6} + \frac{1}{12} + \frac{1}{20} +} \frac{1}{30} + \cdots = & F - \frac{1}{20} & = \frac{1}{5}
\end{array}$$

Summing the first and second columns, we have, combining like fractions,

$$C + D + E + F + G + \cdots = \frac{1}{2} + \frac{2}{6} + \frac{3}{12} + \frac{4}{20} + \frac{5}{30} + \cdots = \sum_{n=2}^{\infty} \frac{1}{n}.$$

On the other hand, summing the first and fourth columns gives

$$C + D + E + F + G + \cdots = 1 + \frac{1}{2} + \frac{1}{3} + \frac{1}{4} + \frac{1}{5} + \cdots = \sum_{n=1}^{\infty} \frac{1}{n}.$$

All the manipulations we have made so far are perfectly justified, where we interpret the summation as an infinite series and acknowledge the possibility that we may have infinite sums (all the appropriate terms are positive, so there cannot be divergence by oscillation). Since these two values of $C + D + E + F + G + \cdots$ differ by 1, we exclude the finite case. Thus the harmonic series diverges.

Proof 9. Recall the notion of a Cauchy sequence from Chapter 1 (Definition 1.45, Theorem 1.46). Let $\{S_n\}$ be the sequence of partial sums of the harmonic series, and assume for the sake of contradiction that the harmonic series converges, so that $\{S_n\}$ is Cauchy. Choose N so that for all $m, n \geq N$ we have $|S_m - S_n| < 1/2$. But for $m = 2N, n = N$,

$$|S_{2N} - S_N| = \sum_{k=N+1}^{2N} \frac{1}{k} \geq \sum_{k=N+1}^{2N} \frac{1}{2N} = \frac{N}{2N} = \frac{1}{2},$$

a contradiction.

Proof 10 [AMM, 107:7, p. 651]. We recall from calculus the inequality

$$x \geq \log(1 + x)$$

that holds for all $x > -1$. This can be proven, for example, by considering the function $f(x) = x - \log(1 + x)$. Since $f'(x) = 1 - 1/(x + 1)$, we see that f is decreasing on $(-1, 0)$ and increasing on $(0, \infty)$; that is, f has an absolute minimum at 0. That is,

$$x - \log(1 + x) \geq 0 - \log(1 + 0) = 0.$$

Using this, we can estimate the partial sums of the harmonic series as follows.

$$\sum_{k=1}^{n} \frac{1}{k} \geq \sum_{k=1}^{n} \log\left(1 + \frac{1}{k}\right) = \sum_{k=1}^{n} \log\left(\frac{k+1}{k}\right) = \sum_{k=1}^{n} (\log(k + 1) - \log k) = \log(n + 1).$$

However, $\log(n + 1)$ goes to infinity for large n, so the partial sums of the infinite series are unbounded, and the harmonic series is divergent.

We can interpret this argument in a much more strikingly visual way as follows. Consider the following graph of the function $g(x) = \sin(\pi e^x)$, shown below. We consider g as a function only of positive reals. We know that this function is defined for arbitrarily large x. We also know that $\sin x$ is zero at integer multiples of π, so that g has zeros whenever e^x is integer-valued, which happens of course for x of the form $\log n$. The distance between consecutive zeros is of the form $\log(k + 1) - \log k$, which by the argument above is a lower bound to $1/k$. This is the motivation for the choice of the function g—the oscillations make visible the segments between zeros, and the lengths of these segments estimate the terms of the harmonic series. If the harmonic series were to converge to some number N, then the length sum of all the segments between zeros of g, since they are smaller, would also be bounded above by N. Then g could have no further zeros right of the vertical line $x = N$, but we know this does not happen. Again we emphasize that this contains no mathematical content not present in the argument above, only a new way to make it tangible.

Proof 11. In Chapter 4, we show that if $\{a_n\}$ is monotone decreasing and $\sum_{n=1}^{\infty} a_n$ converges, then in fact $\lim_{n\to\infty} na_n = 0$. (This is Gem 17.) If the harmonic series were to

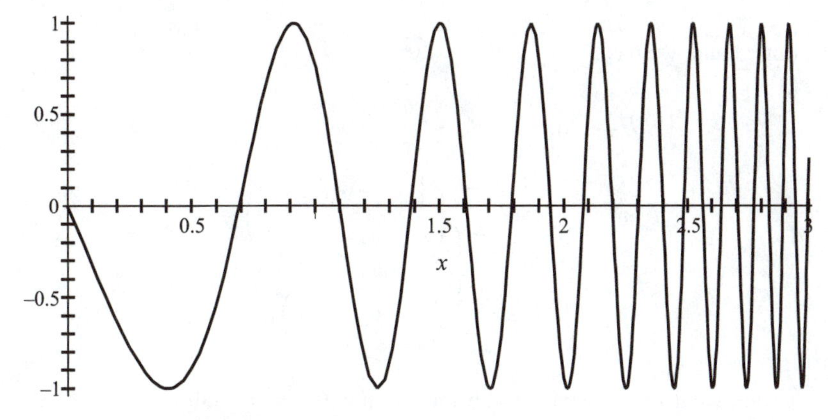

FIGURE 3.1
Visualizing the Divergence of the Harmonic Series

converge, then this gem would apply with $a_n = 1/n$. But in that case, $na_n = 1$ for all n, and it certainly does not converge to 0.

3.2 Rate of Growth

In this section we consider the question of how quickly the harmonic series diverges—more precisely, we ask how quickly the partial sums tend to infinity.

Definition 3.2 *The partial sums of the harmonic series are called the harmonic numbers and they are denoted by* H_n.

Theorem 3.3

$$H_n = \log n + \frac{1}{n} + \gamma_n,$$

where $\{\gamma_n\}$ *is a sequence converging to a constant* γ.

Proof. As in the proof of the Integral Test, we see that

$$\sum_{k=1}^{n-1} \frac{1}{k} \geq \int_1^n \frac{1}{x} dx \geq \sum_{k=2}^{n} \frac{1}{k}.$$

Rewriting this, we have

$$H_{n-1} \geq \log n \geq H_n - 1.$$

Thus from the right inequality we have

$$H_n - \log n \leq 1.$$

On the other hand, we have

$$\begin{aligned} H_n - \log n &= H_n - \int_1^n \frac{1}{x} dx \\ &= \sum_{k=1}^{n-1} \left(\frac{1}{k} - \int_k^{k+1} \frac{1}{x} dx \right) + \frac{1}{n} \\ &= \sum_{k=1}^{n-1} \int_k^{k+1} \left(\frac{1}{k} - \frac{1}{x} \right) dx + \frac{1}{n}. \end{aligned}$$

Since the integrand is nonnegative, the final summation is actually the $(n-1)$th partial sum of a positive series. However, this summation is bounded above by 1 (using the prior inequality), so that it in fact tends to a limit as n goes to ∞. Trivially $1/n$ approaches the limit 0. Then, writing

$$\gamma_n = \sum_{k=1}^{n-1} \int_k^{k+1} \left(\frac{1}{k} - \frac{1}{x} \right) dx,$$

the conclusion follows. Please consult the accompanying figure if this is unclear.

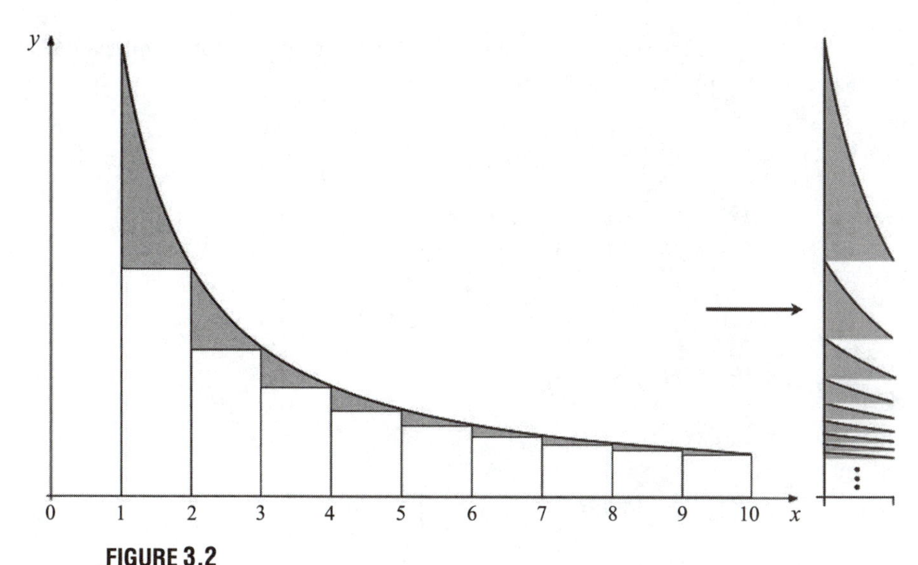

FIGURE 3.2
Estimating Partial Sums of the Harmonic Series; Viewing Gamma as an Area

The above result implies that the harmonic series diverges very slowly indeed with growth comparable to that of $\log n$. The number of terms required to reach a partial sum of A grows exponentially with A. For example, we know that $H_n > 100$ for sufficiently large n. However, we would by no means want to add terms by hand until we reached even a sum as large as 20. It can be checked with computer calculation, for example, that it requires more than a quarter of a billion terms to reach a sum of 20. To reach a partial sum larger than 100 requires approximately 1.5×10^{43} terms.

3.3 The Alternating Harmonic Series

In this section we consider a close relative of the harmonic series.

Definition 3.4 *The alternating harmonic series is the series*

$$\sum_{n=1}^{\infty} \frac{(-1)^{n+1}}{n}.$$

Theorem 3.5 *The alternating harmonic series converges.*

Proof. Apply the Alternating Series Test.

Although we said in Chapter 1 that it is in general extremely difficult to find the exact sum of a convergent series, we can in fact find the sum of the alternating harmonic series.

Theorem 3.6 *The alternating harmonic series converges to* log 2.

Proof 1. For this proof, we change notation and write H_n (resp. A_n) for the nth partial sum of the harmonic (resp. alternating harmonic) series.

$$\begin{aligned}
A_{2n} &= \frac{1}{1} - \frac{1}{2} + \frac{1}{3} - \frac{1}{4} + \frac{1}{5} - \frac{1}{6} + \cdots + \frac{1}{2n-1} - \frac{1}{2n} \\
&= \frac{1}{1} + \left(\frac{1}{2} - \frac{1}{1}\right) + \frac{1}{3} + \left(\frac{1}{4} - \frac{1}{2}\right) + \cdots + \frac{1}{2n-1} + \left(\frac{1}{2n} - \frac{1}{n}\right) \\
&= \left(\frac{1}{1} + \frac{1}{2} + \cdots + \frac{1}{2n}\right) - \left(\frac{1}{1} + \frac{1}{2} + \cdots + \frac{1}{n}\right) \\
&= H_{2n} - H_n.
\end{aligned}$$

Using Theorem 3.3, we have

$$\begin{aligned}
A_{2n} &= \left(\log(2n) + \gamma_{2n} + \frac{1}{2n}\right) - \left(\log n + \gamma_n + \frac{1}{n}\right) \\
&= (\log(2n) - \log n) + (\gamma_{2n} - \gamma_n) - \frac{1}{2n} \\
&= \log 2 + (\gamma_{2n} - \gamma_n) - \frac{1}{2n}.
\end{aligned}$$

Taking the limit, we have

$$\lim_{n\to\infty} A_{2n} = \log 2.$$

Thus the even partial sums of the alternating harmonic series tend to the desired limit. Notice that the difference between an odd partial sum and the nearest even partial sum is tending to zero (since the terms go to 0), so that in fact all partial sums go to log 2, as desired. (We leave the details of this final observation to the reader.)

Proof 2. For this proof we assume a basic knowledge of Taylor Series (see any introductory calculus text). In particular we recall that the function $\log(1+x)$ has the Taylor Series

$$x - \frac{x^2}{2} + \frac{x^3}{3} - \frac{x^4}{4} + \cdots = \sum_{n=1}^{\infty} \frac{(-1)^{n+1}x^n}{n},$$

and that the expansion is valid on the interval $(-1, 1)$. Notice that in effect we are trying to extend this result to $x = 1$. However, standard Taylor Series results only apply to the interior of the interval of convergence, so that it is not straightforward. In particular we cannot just set $x = 1$ and expect anything meaningful. We write

$$s_k(x) = \sum_{n=1}^{k} \frac{(-1)^{n+1} x^n}{n}$$

for the partial sums, where we explicitly acknowledge the dependence on x. We first do some algebra to obtain

$$\begin{aligned} s_k(1) - s_k(x) &= \sum_{n=1}^{k} \frac{(-1)^{n+1}}{n} - \sum_{n=1}^{k} \frac{(-1)^{n+1} x^n}{n} \\ &= \sum_{n=1}^{k} \frac{(-1)^{n+1}(1 - x^n)}{n} \\ &= (1-x) \sum_{n=1}^{k} \frac{(-1)^{n+1}(1 + x + x^2 + \cdots + x^{n-1})}{n}. \end{aligned}$$

Let us examine the summation in the last line. It is in fact the kth partial sum of

$$\sum_{n=1}^{\infty} \frac{(-1)^{n+1}(1 + x + x^2 + \cdots + x^{n-1})}{n}.$$

Restrict our attention now to $x \in (0, 1)$. Then the above series is alternating. Furthermore, we have

$$\begin{aligned} 0 < x < 1 &\Rightarrow x^{n-1} < x^{n-2} < \cdots < x < 1 \\ &\Rightarrow (n-1)x^{n-1} < 1 + x + \cdots + x^{n-2} \\ &\Rightarrow (n-1)(1 + x + \cdots + x^{n-1}) < n(1 + x + \cdots + x^{n-2}) \\ &\Rightarrow \frac{(1 + x + x^2 + \cdots + x^{n-1})}{n} < \frac{(1 + x + x^2 + \cdots + x^{n-2})}{n-1} \end{aligned}$$

so that the terms in the series are decreasing in absolute value. Then the proof of Theorem 1.75 guarantees that partial sums are bounded above by the first term and bounded below by the second partial sum.

$$0 < 1 - \frac{1+x}{2} < \sum_{n=1}^{k} \frac{(-1)^{n+1}(1 + x + x^2 + \cdots + x^{n-1})}{n} < 1.$$

Then

$$0 < s_k(1) - s_k(x) < 1 - x$$

or more simply

$$s_k(x) < s_k(1) < s_k(x) + 1 - x.$$

Passing to the limit as $k \to \infty$, the above inequality will hold, possibly losing strictness, and

$$\log(1+x) \leq \sum_{n=1}^{\infty} \frac{(-1)^{n+1}}{n} \leq \log(1+x) + 1 - x.$$

Since these bounds on our sum hold for all $x \in (0, 1)$, they hold also in the limit as $x \to 1$, giving

$$\log 2 \leq \sum_{n=1}^{\infty} \frac{(-1)^{n+1}}{n} \leq \log 2$$

and we are done.

Theorem 3.7 *The alternating harmonic series can be rearranged to sum to any prescribed real number r. It can also be rearranged to diverge to ∞ or to $-\infty$ as well as to diverge by oscillation.*

Proof. By the reasoning of Proof 3 that the harmonic series diverges, we see that the series consisting only of the positive terms in the alternating harmonic series is also divergent, and likewise with the negative terms. That is, no matter how many terms we have included from either class, there will always remain enough positive (or negative) terms to increase (decrease) the partial sums of the rearranged series as much as desired. Notice that we are not using the extent of our freedom in this result, since the rearrangement described in the following leave all the positive (resp. negative) terms in the same relative order. That is, we will always take the positive terms in order and likewise negative terms, but we will change the pattern of interweaving them.

Let r be a real number. To form our rearrangement, we add positive terms until the partial sum exceeds r for the first time. We then add negative terms until the partial sum becomes less than r. We then add positive terms until the partial sum becomes greater than r, and so on. We leave to the reader to check that all terms must eventually be used. Now, it remains only to show that the series converges to r. Consider the places where the sequence of partial sums "changes direction." Since each positive (resp. negative) term is less in absolute value than all those before it, it can be verified that the "overshoots" get smaller and smaller, and likewise the "undershoots." It follows that the partial sums do indeed approach r.

To effect divergence to ∞, we add positive terms until the sum passes 1, then a single negative term. We add positive terms until the sum passes 2, then a single negative term. We add positive terms until the sum passes 3, then a single negative term, and so on. It is clear that this series diverges to ∞. A similar strategy would solve the case of $-\infty$.

To effect divergence by oscillation, we add positive terms until the sum passes 1, then negative terms until the sum passes -2, then positive terms until the sum passes 3, and so on.

Theorem 3.8 *Let $\sum_{n=1}^{\infty} a_n$ be a series with terms going to 0 and such that the subseries consisting of only positive or only negative terms would themselves diverge. This series* *can be rearranged to sum to any prescribed real number r. It can also be rearranged to diverge to ∞ or to $-\infty$ as well as to diverge by oscillation.*

Proof. The proof follows the pattern of the special case already proven. We leave the details to the reader.

The following result shows that this is a special case of a more general result. In essence the following result says, in most cases, that either rearrangement has no effect on a series or it can have any conceivable effect on the series. Intermediate cases do not occur.

Theorem 3.9 *Let $\sum_{n=1}^{\infty} a_n$ be a series whose terms go to 0. Then exactly one of the following holds.*

- *$\sum_{n=1}^{\infty} a_n$ converges absolutely and all its rearrangements have the same sum.*
- *All rearrangements of the series diverge to ∞.*
- *All rearrangements of the series diverge to $-\infty$.*
- *The series can be rearranged to sum to any prescribed real number r. It can also be rearranged to diverge to ∞ or to $-\infty$ as well as to diverge by oscillation.*

Proof. Consider two subseries of $\sum_{n=1}^{\infty} a_n$, the one consisting of exactly the positive terms and the one consisting of exactly the negative terms, with the understanding that one (but not both) of these might be finite. Now *a priori* either or both may converge. The reader can verify after a moment's thought that the last case will occur if both diverge by Theorem 3.8, the second will occur if only the positive subseries converges, and the third will occur if only the negative subseries converges. Furthermore, if both subseries converge, then $\sum_{n=1}^{\infty} |a_n|$ converges to the sum of the absolute values of their sums, so the series is absolutely convergent, and we are in the first case.

We demonstrate the possibilities first with a single example.

Example 3.10 (compare Putnam 1954, Afternoon 2) *The series* $1+1/3-1/2+1/5+1/7-1/4+1/9+1/11-1/6+\cdots$ *converges to* $\frac{3}{2}\log 2$.

Solution. Let s_n be the nth partial sum of the alternating harmonic series, and let t_n be the nth partial sum of the above rearrangement of the alternating harmonic series. We make the following manipulation.

$$\begin{aligned}
s_{4n} &= \frac{1}{1} - \frac{1}{2} + \frac{1}{3} - \frac{1}{4} + \cdots + \frac{1}{4n-3} - \frac{1}{4n-2} + \frac{1}{4n-1} - \frac{1}{4n} \\
\frac{1}{2}s_{2n} &= \phantom{\frac{1}{1} -{}} \frac{1}{2} - \phantom{\frac{1}{3}} \frac{1}{4} + \cdots \phantom{\frac{1}{4n-3}} + \frac{1}{4n-2} \phantom{\frac{1}{4n-1}} - \frac{1}{4n}. \\
s_{4n} + \frac{1}{2}s_{2n} &= \frac{1}{1} + \phantom{\frac{1}{2}} \frac{1}{3} - \frac{1}{2} + \cdots + \frac{1}{4n-3} \phantom{\frac{1}{4n-2}} + \frac{1}{4n-1} - \frac{1}{2n}
\end{aligned}$$

That is, $t_{3n} = s_{4n} + \frac{1}{2}s_{2n}$. By adding individual terms, we have

$$t_{3n} = s_{4n} + \frac{1}{2}s_{2n}$$

$$t_{3n+1} = s_{4n} + \frac{1}{2}s_{2n} + \frac{1}{4n+1}$$

$$t_{3n+2} = s_{4n} + \frac{1}{2}s_{2n} + \frac{1}{4n+1} + \frac{1}{4n+3}.$$

As n goes to ∞, we see that all three right-hand sides tend to $\log 2 + \frac{1}{2}\log 2 = \frac{3}{2}\log 2$. Thus t_n also approaches this limit, and we are done.

Proof 12 (This is the final proof that the harmonic series diverges, as promised at the beginning of the chapter.) Notice that Example 3.10 gives a rearrangement of the alternating harmonic series with a different sum by explicit construction. By Theorem 1.71, the alternating harmonic series must converge conditionally. Since the alternating harmonic series does not converge absolutely, the harmonic series diverges.

Example 3.10 can actually be generalized considerably. For the generalization we use a proof that is more technical, relying on the technical proof of Theorem 3.3 rather than a slick manipulation.

Theorem 3.11 *Consider the rearrangement of the alternating harmonic series consisting of p positive terms, then q negative terms, then p positive terms, then q negative terms, etc., where the positive (resp. negative) terms stay in the same relative order. (Example 3.10 is the special case $p = 2, q = 1$.) Then this series converges. Furthermore, its sum is* $\log 2 + \frac{1}{2}(\log p - \log q)$.

Proof. We use Theorem 3.3. Let s_n be the partial sums of this rearrangement. Consider only partial sums that consist of entire "blocks" of terms. (That is, since the pattern of positive and negative terms has period $p+q$, we consider the partial sums $s_{(p+q)n}$. Since the sum for each block tends to 0 (an easy exercise), this is sufficient to demonstrate convergence of the series. We simply write

$$\begin{aligned} s_{(p+q)n} &= \left(\frac{1}{1} + \frac{1}{3} + \cdots + \frac{1}{2p-1} - \frac{1}{2} - \frac{1}{4} - \cdots - \frac{1}{2q}\right) + \cdots \\ &\quad + \left(\frac{1}{2p(n-1)+1} + \frac{1}{2p(n-1)+3} + \cdots \right. \\ &\quad \left. + \frac{1}{2pn-1} - \frac{1}{2q(n-1)+2} - \frac{1}{2q(n-2)+4} - \cdots - \frac{1}{2qn}\right) \\ &= \left(\frac{1}{1} + \frac{1}{3} + \cdots + \frac{1}{2pn-1}\right) - \left(\frac{1}{2} + \frac{1}{4} + \cdots + \frac{1}{2qn}\right) \end{aligned}$$

$$= \left(\frac{1}{1} + \frac{1}{2} + \frac{1}{3} + \cdots + \frac{1}{2pn}\right) - \left(\frac{1}{2} + \frac{1}{4} + \cdots + \frac{1}{2pn}\right)$$
$$- \left(\frac{1}{2} + \frac{1}{4} + \cdots + \frac{1}{2qn}\right)$$
$$= H_{2pn} - \frac{1}{2}H_{pn} - \frac{1}{2}H_{qn}$$
$$= \log(2pn) + \gamma_{2pn} + \frac{1}{2pn} - \frac{1}{2}\log(pn) - \frac{1}{2}\gamma_{pn} - \frac{1}{2pn}$$
$$-\frac{1}{2}\log(qn) - \frac{1}{2}\gamma_{qn} - \frac{1}{2qn}$$
$$= \log 2 + \log p + \log n - \frac{1}{2}\log p - \frac{1}{2}\log n - \frac{1}{2}\log q - \frac{1}{2}\log n$$
$$+ \gamma_{2pn} - \frac{1}{2}\gamma_{pn} - \frac{1}{2}\gamma_{qn} - \frac{1}{2qn}$$
$$= \log 2 + \frac{1}{2}(\log p - \log q) + \gamma_{2pn} - \frac{1}{2}\gamma_{pn} - \frac{1}{2}\gamma_{qn} - \frac{1}{2qn}.$$

In the limit, the γ_i terms cancel and the $1/2qn$ vanishes, and the desired result remains.

The interested reader can learn more about the alternating harmonic series and its rearrangements by studying *CMJ*, Vol. 16 pp. 135–138; *AMM* Vol. 87 pp. 817–819; *AMM* Vol. 88 pp. 33–46; *MM* Vol. 54 pp. 244–246.

3.4 Selective Sums

In this section, we deal with the concept of selective sums. Loosely speaking, we consider the possibilities that result when we add only a subset of the terms in the series. We are more explicit in the following definitions.

Definition 3.12 *Let $\sum_{n=1}^{\infty} a_n$ be an infinite series. Then a number r is a selective sum of this series if we can write $r = \sum_{n \in A} a_n$ for some suitably chosen nonempty subset A of the positive integers.*

In the first subsection, we demonstrate that all positive numbers are in fact selective sums of the harmonic series, which is a considerably stronger result than its mere divergence. In the latter two subsections, we consider two particular selective sums of the harmonic series and investigate their convergence or divergence.

Existence Theorems

Theorem 3.13 *Every positive number r is a selective sum for the harmonic series; furthermore the set of terms chosen can always be chosen infinite.*

Proof. We apply the Greedy Algorithm, at each stage taking the largest possible next fraction. Explicitly, we construct our set recursively as follows, placing k into the set A if

$$\left(\sum_{n\in A;n<k}\frac{1}{n}\right)+\frac{1}{k}<r.$$

Intuitively, we include a term if it is possible to do so subject to the constraint that all partial sums stay below our target r. Since our series consists of positive terms and all partial sums are bounded by r, the series must converge to a sum $s\leq r$. In fact we claim that $s=r$. Suppose for the sake of contradiction that $s<r$, and choose N sufficiently large that $1/N<r-s$. Then adding $1/N$ would not have caused us to reach or exceed r, so that $1/N$ must have been included in our summation. But this reasoning applies to all N sufficiently large. That is, for all N such that $1/N<r-s$, $1/N$ must already be in our summation. Then our summation must include a tail of the harmonic series. Since the harmonic series diverges, so must any tail and hence our series. But our series has bounded partial sums, a contradiction.

By construction, in the previous theorem the partial sums never actually equaled our target r, so that our selective sums originated from genuinely infinite series. One wonders under what circumstances we could have achieved our target sum with a finite series. Certainly the sum of a finite number of fractions would be a rational number. The following theorem shows that this constraint is also sufficient.

Theorem 3.14 (compare Putnam Problem [14, 1954, B2]) *Every positive rational number r is a selective sum for the harmonic series consisting of finitely many terms.*

Proof. We prove first the case $r<1$, doing this by induction on the numerator of r. If r has numerator 1, then r is itself a term in the harmonic series and we are done.

Suppose that the conclusion holds for all numbers r with numerator less than p, and let $r=p/q$. Choose the largest possible fraction not exceeding r, so that

$$\frac{1}{n}\leq\frac{p}{q}<\frac{1}{n-1}.$$

By the second inequality, $p(n-1)<q\Rightarrow pn-q<p$. However,

$$\frac{p}{q}-\frac{1}{n}=\frac{pn-q}{qn},$$

so that the remainder after including $1/n$ into the sum has numerator less than p. By the induction hypothesis, we can write this remainder as a finite sum of unit fractions (that is to say, fractions with numerator 1). Furthermore, since

$$\frac{p}{q}-\frac{1}{n}<\frac{1}{n-1}-\frac{1}{n}=\frac{1}{n(n-1)}\leq\frac{1}{n},$$

the terms required to obtain $p/q-1/n$ must not include $1/n$. Then appending $1/n$ to the sum for $p/q-1/n$ gives the desired sum for r, completing the induction.

If $r \geq 1$, then consider the largest partial sum of the harmonic series not exceeding r. That is, find N so that

$$\sum_{n=1}^{N} \frac{1}{n} \leq r < \sum_{n=1}^{N+1} \frac{1}{n},$$

and include the first N terms of the harmonic series in our sum. The remainder, if any, is therefore less than $1/N + 1$. By the above result it can be written with a finite number of unit fractions, and it is clear that they are sufficiently small not to have been previously included.

In any case, then, we can construct a finite series as desired.

We note that the above finite expansion of a rational number into unit fractions was the notation for rationals used in ancient Egypt. For this reason, forms like $1/2 + 1/3 + 1/19$ are called Egyptian fractions.

Example 3.15 *Express $2/5$ as a finite selective sum of the harmonic series (i.e., as an Egyptian fraction).*

Solution. Solving

$$\frac{1}{n} \leq \frac{2}{5} < \frac{1}{n-1},$$

we see our first fraction should have $n = 3$. Now $2/5 - 1/3 = 1/15$. Hence $2/5 = 1/3 + 1/15$.

In fact we have much more freedom in choosing an expression, both in the infinite and finite case, than we are using. Since any tail of the harmonic series diverges, we can restrict our attention to such tails.

Theorem 3.16 *Let N be an arbitrary positive integer. Then every positive number r is realizable as a selective sum of the harmonic series without using any of the first N terms of the harmonic series. If r is rational, the selective sum can be realized with finitely many terms.*

Proof. Essentially identical to the given proofs. The details are left to the reader.

Casting Out Nines

Suppose that we remove from the harmonic series all terms that contain a 9 in their denominator. At first it seems that this cannot make too much difference in the sum of the series. Near the beginning of the series, after all, we are only removing every tenth term. Strikingly, however, this change is enough to make the series converge.

Theorem 3.17 *Let B be the set of positive integers whose decimal representations do not contain the digit 9. Then the series*

$$\sum_{n\in B}\frac{1}{n}$$

converges.

Proof. Let M_d be the contribution to the sum from terms in which the denominator is a d-digit number. For example,

$$M_1=\frac{1}{1}+\frac{1}{2}+\frac{1}{3}+\frac{1}{4}+\frac{1}{5}+\frac{1}{6}+\frac{1}{7}+\frac{1}{8}\le 8.$$

To estimate M_2, notice that there are 8 possible first digits for a two-digit number in B and 9 possible second digits. That is, there will be 72 terms with a two-digit denominator. Since each term is at most $1/10$, the total contribution is at most $72/10$.

More generally, consider the d-digit numbers in B. There are 8 ways to choose the first digit and 9 ways to choose each of the remaining $d-1$ digits. That is, there will be $8(9)^{d-1}$ terms, each at most $1/10^{d-1}$. Totaling these, we have

$$M_d\le 8(9)^{d-1}\left(\frac{1}{10^{d-1}}\right)=8\left(\frac{9}{10}\right)^{d-1}.$$

We now notice that

$$\sum_{n\in B}\frac{1}{n}=M_1+M_2+M_3+\cdots\le\sum_{d=1}^{\infty}8\left(\frac{9}{10}\right)^{d-1}.$$

The infinite series on the right, however, is a convergent geometric series with common ratio $9/10$, so that this thinned harmonic series converges (perhaps unexpectedly).

This result is perhaps not so discordant with common sense if we examine it from a different perspective. The authors were deliberately manipulative in directing the reader's attention to the beginning of the series and the relative scarcity of numbers with 9's. Rather, recall that the end behavior of a series is determined by its tail, not its initial segment. If we look at numbers with, say, 100 digits or more, it is clear that nearly all of them will have 9's and thus will be excluded from the sum. We leave it to the reader to verify that the same result is true if the digit 9 is replaced by any other digit. In fact much more is true, as stated in the following theorem (the proof of which we omit).

Theorem 3.18 *Let A be the set of positive integers whose decimal representations contain at least a_1 1's, at least a_2 2's, at least a_3 3's ..., at least a_9 9's, and at least a_0 0's, and let B be the complement of A in the set of positive integers. Then the series*

$$\sum_{n\in B}\frac{1}{n}$$

converges.

Proof. We direct the reader to a selection of articles on these "thinned" harmonic series. See *The American Mathematical Monthly*, for example, Vol. 20 (pp. 48–50); Vol. 23 (pp. 149–152); Vol. 23 (pp. 302–303); Vol. 48 (pp. 93–97); Vol. 82 (pp. 931–933).

Prime Harmonic Series

Definition 3.19 *An infinite product is an expression of the form*

$$\prod_{n=1}^{\infty} a_n,$$

where $\{a_n\}$ is a sequence of positive numbers called factors. To this infinite product we associate the sequence of partial products

$$P_k = \prod_{n=1}^{k} a_n.$$

As with infinite series, we say that the infinite product converges if the sequence of partial products converges. Otherwise, we say the product diverges.

Theorem 3.20 *Let $\{a_n\}$ be a sequence of positive numbers. Then the infinite product $\prod_{n=1}^{\infty}(1+a_n)$ and the infinite sum $\sum_{n=1}^{\infty} a_n$ converge or diverge together.*

Proof. We will show that the convergence of either implies the convergence of the other. Denote the partial products and partial sums as follows:

$$P_k = \prod_{n=1}^{k}(1+a_n)$$

$$S_k = \sum_{n=1}^{k} a_n.$$

Now, it is clear that the sequences $\{P_k\}$ and $\{S_k\}$ increase with k, so that they will converge if and only if they are bounded above. We note first that

$$P_k = 1 + a_1 + a_2 + a_3 + \cdots + a_k + a_1 a_2 + \cdots \geq 1 + a_1 + a_2 + a_3 + \cdots + a_k = 1 + S_k.$$

We also recall again from calculus the inequality

$$\log(1+x) \leq x$$

which holds for all $x > -1$.

With this in mind, we have

$$\log P_k = \sum_{n=1}^{k} \log(1+a_n) \leq \sum_{n=1}^{k} a_n = S_k.$$

Combining these, we have

$$S_k + 1 \le P_k \le e^{S_k}.$$

It follows that, if one of the sequences is bounded above, the other must be also.

Theorem 3.21 *Let p_i be the ith prime number. (That is, $p_1 = 2$, $p_2 = 3$, $p_3 = 5$, $p_4 = 7, \ldots$.) Then*

$$\sum_{i=1}^{\infty} \frac{1}{p_i}$$

diverges.

Proof. Recall the Fundamental Theorem of Arithmetic, which says that each positive integer has a unique factorization into prime powers. Operating formally, if we expand the infinite product

$$P = \left(1 + \frac{1}{2} + \frac{1}{4} + \cdots\right)\left(1 + \frac{1}{3} + \frac{1}{9} + \cdots\right) \cdots \left(1 + \frac{1}{p_i} + \frac{1}{p_i^2} + \cdots\right),$$

we will obtain each unit fraction $1/n$ exactly once, so that by a rearrangement (legitimate because all the terms are positive)

$$P = \sum_{n=1}^{\infty} \frac{1}{n},$$

so that in fact P diverges as an infinite product. On the other hand, each factor is a geometric series, and we can use Theorem 1.39 to write

$$\begin{aligned} P &= \prod_{i=1}^{\infty}\left(1 + \frac{1}{p_i} + \frac{1}{p_i^2} + \cdots\right) = \prod_{i=1}^{\infty}\left(\frac{1}{1 - \frac{1}{p_i}}\right) \\ &= \prod_{i=1}^{\infty}\left(\frac{p_i}{p_i - 1}\right) = \prod_{i=1}^{\infty}\left(1 + \frac{1}{p_i - 1}\right). \end{aligned}$$

We know therefore that this product diverges. Using Theorem 3.20, we know also that

$$\sum_{i=1}^{\infty} \frac{1}{p_i - 1}$$

diverges.

Finally, we use the Limit Comparison Test to relate this to the prime harmonic series. We simply note that

$$\lim_{i\to\infty} \frac{1/p_i}{1/(p_i - 1)} = \frac{p_i - 1}{p_i} = 1,$$

so that the prime harmonic series diverges as desired.

This result is interesting to number theorists because it gives some information about how the primes are distributed throughout the natural numbers. Loosely speaking, primes must be relatively common for the above subseries to diverge. We will settle here for the coarsest possible result.

Corollary 3.21.1 *There are an infinite number of primes.*

3.5 Unexpected Appearances

The harmonic series has a surprising habit of rearing its head in a wide variety of problems throughout mathematics. In this book we include three such examples of intriguing problems of the sort that could be given to a math student for exploration. (The third is elsewhere, and we leave it to the reader to find.) Of course, the reader will be less surprised than the authors were to find the harmonic series making an entrance in the following problems, since they appear neatly grouped in this chapter about harmonic series. Still, in each case, the reader will find that the divergence of the harmonic series plays a key role in the solution in forcing the (often counterintuitive) answer to the question. It is easy to just hear the sentence "The harmonic series diverges." without getting excited, because it is a technical sentence. However, the following results are often challenging to common sense. The reader can consider these problems as being concrete realizations of the bizarreness of the divergence of the harmonic series.

Example 3.22 *Lucky the Worm is trying to cross an elastic rope. The rope is one foot long, and Lucky can crawl at a steady pace of one inch in a minute. However, at the end of each minute, Lucky takes a short breather while the rope is stretched an additional foot. We assume that the stretching is done uniformly across the rope, so that Lucky is the same fraction of the way along the rope both before and after the stretching. We also assume that Lucky himself has negligible length, and that he survives the stretchings (he's Lucky after all). Will Lucky be able to reach the far end of the rope?*

Solution. At first glance, things don't look very good for Lucky. Length is being added to the rope at twelve times the rate that he is covering distance, so it doesn't seem like he could possibly progress much. However, we will show that in fact Lucky could cross the rope. In the first minute, Lucky will cover one inch of a foot long rope, accomplishing $1/12$ of his goal. In the second minute, Lucky will cover one inch out of a two foot rope, accomplishing $1/24$ of his goal. In the third minute, Lucky will cover one inch out of a three foot rope, accomplishing $1/36$ of his goal. The total progress in n minutes, then, is

$$\frac{1}{12} + \frac{1}{24} + \frac{1}{36} + \cdots + \frac{1}{12n} = \frac{1}{12}\left(\frac{1}{1} + \frac{1}{2} + \frac{1}{3} + \cdots + \frac{1}{n}\right).$$

We recognize the harmonic series in the above. Since the harmonic series diverges, we can choose n large enough that the partial sum of the harmonic series exceeds 12. Lucky's total progress in n minutes, then, would reach the desired total of 1, that is 100% of the task, or the entire rope.

We wish to emphasize the counterintuitive nature of this example, possibly straining the reader's credulity further. Notice that there was nothing special about the number 12. If Lucky were trying to cross a mile-long rope that grew by a mile every minute, still crossing at a lowly inch per minute, he would be able to accomplish that as well. (The interested reader may wish to use Theorem 3.3 to estimate how long each of these tasks would take Lucky.)

Example 3.23 *Is it possible to stack dominoes on a table in such a way that the top domino totally overhangs the table?*

Solution. Without loss of generality suppose that the dominoes have length 2 units, so that each edge is 1 unit from the center of the domino. We claim that, in a stack of n dominoes, the top domino can be placed so that its rightmost edge is H_n units beyond the edge of the table. Choosing n sufficiently large, we will have $H_n > 2$ as desired. In fact we can arrange things so that the top domino is as far away from the table as the patient stacker may wish.

To do this, we let the edge of the table be the origin of our number line, and number the dominoes from the top down so that domino n rests on the table. We place domino n so that it overhangs the table by $1/n$ units, and thereafter each domino k overhangs domino $k+1$ by $1/k$ units. By construction, domino k will have its rightmost edge at

$$r_k = \sum_{i=k}^{n} \frac{1}{i}$$

and its center at

$$c_k = r_k - 1 = \sum_{i=k}^{n} \frac{1}{i} - 1.$$

It is clear then that the right edge of the top domino will be at H_n as desired. What is unclear is whether the domino stack will fall. The domino stack will balance if the center of mass of the top k dominoes is over the $k+1$st domino, for every k (where the $(n+1)$st domino is in fact the table). The horizontal coordinate of the center of mass of several dominoes is just the average of the coordinates of the centers of mass of individual dominoes. Combining these observations, the dominoes will balance if and only if for all k,

$$r_{k+1} \geq \frac{1}{k} \sum_{n=1}^{k} c_n,$$

where we put

$$r_{n+1} = \sum_{i=n+1}^{n} \frac{1}{i} = 0.$$

We simply verify

$$\begin{aligned}
\frac{1}{k}\sum_{j=1}^{k} c_j &= \frac{1}{k}\sum_{j=1}^{k}\left(\sum_{i=j}^{n}\frac{1}{i}-1\right)\\
&= \frac{1}{k}\left(\sum_{j=1}^{k}\sum_{i=j}^{n}\frac{1}{i}\right)-1\\
&= \frac{1}{k}\sum_{j=1}^{k}\left(\sum_{i=j}^{k}\frac{1}{i}+\sum_{i=k+1}^{n}\frac{1}{i}\right)-1\\
&= \frac{1}{k}\sum_{j=1}^{k}\sum_{i=j}^{k}\frac{1}{i}+\sum_{i=k+1}^{n}\frac{1}{i}-1\\
&= \frac{1}{k}\sum_{j=1}^{k}\sum_{i=j}^{k}\frac{1}{i}-1+r_{k+1}.
\end{aligned}$$

Now, in the above double summation, each term $1/i$ appears once each when $j=1$, $j=2$, $j=3,\ldots,j=i$, a total of i times. Then we can write

$$\begin{aligned}
\frac{1}{k}\sum_{j=1}^{k} c_j &= \frac{1}{k}\sum_{j=1}^{k}\sum_{i=j}^{k}\frac{1}{i}-1+r_{k+1}\\
&= \frac{1}{k}\sum_{i=1}^{k} i\left(\frac{1}{i}\right)-1+r_{k+1}\\
&= r_{k+1},
\end{aligned}$$

as desired. The fact that equality holds signals that we have positioned our dominoes optimally in the sense that placing any domino any further from the table will cause a collapse.

The reader can actually do this experiment if he or she desires. We illustrate the situation for six dominoes below. The x-coordinate for the center of mass for each of the dominoes

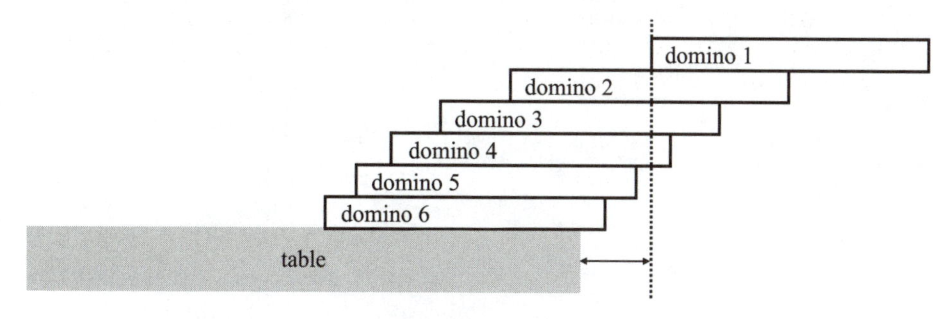

FIGURE 3.3
Stacking Dominoes

is (from bottom to top) 0, $1/5$, $1/5 + 1/4$, $1/5 + 1/4 + 1/3$, $1/5 + 1/4 + 1/3 + 1/2$, $1/5 + 1/4 + 1/3 + 1/2 + 1$. The x coordinate for the center of mass for the entire stack of six dominoes is

$$\frac{5(1/5) + 4(1/4) + 3(1/3) + 2(1/2) + 1}{6} = \frac{5}{6} < 1.$$

That is, the center of mass lies over the last domino, preventing tumbling. We remark that the top domino can be made to wholly overhang the table with as few as four dominoes, if your hand is steady.

4

Intriguing Results

The theory of infinite series is particularly interesting in mathematics due to its wealth of small and interesting results. While there is a definite fabric of results that form the standard theory, there are innumerable results that do not quite fit within a linear development but are no less interesting. For example, we have said earlier in the book that no general method exists for summing a particular convergent series. However, certain special series may be summable by a particularly clever or elegant method. Results of that form can be beautiful as well as instructive, often exhibiting clues to summing other similar series. In this chapter we include a collection of results that we find intriguing for one reason or another on their own merits, not because of their place in the larger fabric. (However, some of these results would certainly have a place in other authors' development of the theory.)

How, then, were results chosen for this chapter? On what basis do we call a result a gem? There are a number of reasons we may have labelled a result a gem. Gems may confirm a common intuitive notion with a clarifying proof, or may provide a counterexample to intuition. They may exhibit a particularly slick or unexpected proof technique. Some gems were chosen because they seemed to us at first astonishing but with a moment's study of the proof became almost common sense. With few exceptions the proofs of these results are elementary and can be appreciated by anyone who has successfully come this far through the book. In a sense that is one purpose of this chapter—to demonstrate that the theory of infinite series is full of surprising and intriguing gems that are above all *accessible* to the nonspecialist.

The reader should not feel any obligation to sit down and read all of these results in a sitting. Quite the contrary, this chapter is designed to be appreciated in bits and pieces, according to the reader's interest. For example, the reader interested in exotic series that can be summed exactly can browse the chapter for such examples. Most of the gems stand alone and apart from the rest of the gems in this chapter, though they may reference earlier chapters. In the cases where one gem depends on another or can be better interpreted in the context of other items in the chapter, this is always explicit. The reader is invited to sample this chapter freely; this chapter is by its very nature not exhaustive, so we welcome the reader to use these results as jumping-off points for intriguing results of his or her own.

4.1 Gems

Gem 1 *Let $\{a_n\}$ be a sequence of terms such that $\sum_{n=1}^{\infty} a_n^2$ converges. Then*

$$\sum_{n=1}^{\infty} \frac{a_n}{n}$$

converges absolutely.

Proof. Recall that $(|a| - |b|)^2 \geq 0$ so that by rearranging we have $|a|^2 + |b|^2 \geq 2|a||b|$. Let a_n be given as above, and let $b_n = 1/n$. Applying this result to a_i, b_i for each $i = 1, 2, \ldots, k$ and summing,

$$\sum_{n=1}^{k} a_n^2 + \sum_{n=1}^{k} \frac{1}{n^2} \geq 2 \sum_{n=1}^{k} \frac{|a_n|}{n}.$$

Since the two series on the left-hand side are convergent series of positive terms, the left-hand side is bounded above by the sum of their sums, independently of k. Thus the partial sums of the series

$$\sum_{n=1}^{\infty} \frac{|a_n|}{n}$$

are bounded above. Since this is a series of positive terms with bounded partial sums, it must converge. Thus

$$\sum_{n=1}^{\infty} \frac{a_n}{n}$$

converges absolutely, as desired. (This could also be done using the Cauchy-Schwartz Inequality.)

Gem 2 *Let $\{a_n\}$, $\{b_n\}$ be two sequences of terms such that $\sum_{n=1}^{\infty} a_n^2$ and $\sum_{n=1}^{\infty} b_n^2$ converge. Then $\sum_{n=1}^{\infty} a_n b_n$ converges absolutely.*[1]

Proof. We follow the proof of Gem 1. Again we have $|a|^2 + |b|^2 \geq 2|a||b|$. Applying this result to a_i, b_i for each $i = 1, 2, \ldots, k$ and summing,

$$\sum_{n=1}^{k} a_n^2 + \sum_{n=1}^{k} b_n^2 \geq 2 \sum_{n=1}^{k} |a_n||b_n|.$$

[1]This relation between the convergence of the "square sums" of series and the convergence of their term-by-term product is important to the study of Hilbert spaces (which we do not discuss here). The sum of the term-by-term product forms what is called an *inner product* on a Hilbert space, and this gem and its generalizations guarantee its existence and describe some of its properties.

Since the two series on the left-hand side are convergent series of positive terms, the left-hand side is bounded above by the sum of their sums, independently of k. Thus the partial sums of the series

$$\sum_{n=1}^{\infty} |a_n||b_n|$$

are bounded above. Since this is a series of positive terms with bounded partial sums, it must converge. Thus $\sum_{n=1}^{\infty} a_n b_n$ converges absolutely, as desired.

Gem 3 *The theory of infinite sequences and the theory of infinite series are logically equivalent in the sense that the behavior of any sequence is controlled by the behavior of an associated series, and vice versa.*

Proof. In defining the convergence of infinite series we established a correspondence from infinite series $\sum_{n=1}^{\infty} a_n$ to infinite sequences of partial sums $\{S_n\}$. We make the observation that, since $S_n = a_1 + a_2 + \cdots + a_{n-1} + a_n = S_{n-1} + a_n$ for $n > 1$, we have $a_n = S_n - S_{n-1}$ for $n > 1$ and also $a_1 = S_1$. That is, the terms $\{a_n\}$ are uniquely determined by the partial sums $\{S_n\}$. In other words, the correspondence from infinite series to infinite sequences is in fact a bijection.

In this development of infinite series, we have followed the standard route of defining infinite sequences and defining the behavior of infinite series in terms of the associated sequence of partial sums. Instead we could have defined the convergence of infinite series as our primitive concept. Then the convergence and divergence of infinite sequences could be defined in terms of the associated series. The next result makes this a bit more explicit.

Gem 4 *Let $\{a_n\}$ be a sequence of real numbers, and define $\{b_n\}$ by $b_n = a_{n+1} - a_n$. Then $\sum_{n=1}^{\infty} b_n$ converges if and only if $\{a_n\}$ converges.*

Proof. We essentially follow the definitions. $\sum_{n=1}^{\infty} b_n$ converges if and only if

$$\sum_{n=1}^{\infty} (a_{n+1} - a_n)$$

converges if and only if $\sum_{n=1}^{k} (a_{n+1} - a_n) = (a_{k+1} - a_1)$ converges if and only if $\{a_{k+1}\}$ converges if and only if $\{a_n\}$ converges.

Gem 5 *If $\sum_{n=1}^{\infty} a_n$ and $\sum_{n=1}^{\infty} b_n$ are convergent series of positive terms, then*

$$\sum_{n=1}^{\infty} \sqrt{a_n^2 + b_n^2}$$

converges also.

Proof. Since the terms are all positive we have

$$a_i^2 + b_i^2 < a_i^2 + 2a_ib_i + b_i^2 = (a_i + b_i)^2,$$

we have, by the Comparison Test,

$$\sum_{n=1}^{\infty} \sqrt{a_n^2 + b_n^2} < \sum_{n=1}^{\infty} a_n + \sum_{n=1}^{\infty} b_n.$$

The two series on the right side converge by hypothesis, so that the series on the left side converges also, as desired.

Gem 6 *If* $\sum_{n=1}^{\infty} \sqrt{a_n^2 + b_n^2}$ *converges, then* $\sum_{n=1}^{\infty} a_n$ *and* $\sum_{n=1}^{\infty} b_n$ *converge absolutely.*

Proof. This converse to Gem 5 is straightforward using the Comparison Test. We simply notice

$$0 \le |a_n|, |b_n| \le \sqrt{a_n^2 + b_n^2},$$

and apply the Comparison Test.

Definition 4.1 *Let I be any set, and let $\{a_i\}_{i \in I}$ be a sequence of real numbers indexed by I. Then we call the expression*

$$\sum_{i \in I} a_i$$

an unordered series.[2] *We say that the series converges to a real number S if and only if for each $\epsilon > 0$ there exists a finite set $J_0 \subset I$ with the property that, for every finite set J satisfying $J_0 \subset J \subset I$, we have*

$$\left| \sum_{i \in J} a_i - S \right| < \epsilon.$$

In this case we say S is the sum of the series. If the series does not converge to any number, it is said to diverge. If for every $N > 0$, there exists a finite set $J_0 \subset I$ with the property that, for every finite set J satisfying $J_0 \subset J \subset I$, we have

$$\sum_{i \in J} a_i > N,$$

we say that the series diverges properly or diverges to ∞.

[2]This notion of unordered series convergence is important in more complicated topological spaces where the ordinary notion of sequence is inadequate. The notion of *net* is closely related.

Gem 7 *Let $\{a_n\}$ be a sequence of real numbers. Then $\sum_{n \in \mathbf{N}} a_n$ converges as an unordered series if and only if $\sum_{n=0}^{\infty} a_n$ converges absolutely, and furthermore the sums are the same. Here $\mathbf{N}$ means the set of natural numbers $\{0, 1, 2, 3, \dots, \}$.*

Proof. First suppose that $\sum_{n=0}^{\infty} a_n$ converges absolutely to a sum S. Let $\epsilon > 0$ be given. Since $\sum_{n=0}^{\infty} |a_n|$ converges, we can find a number N such that

$$\sum_{n=N+1}^{\infty} |a_n| < \epsilon.$$

Let $J_0 = \{0, 1, 2, \dots, N\}$. Then if J is a finite set containing J_0, we have

$$\left| S - \sum_{n \in J} a_n \right| = \left| \sum_{n=0}^{\infty} a_n - \sum_{n \in J} a_n \right| = \left| \sum_{n=N+1}^{\infty} a_n - \sum_{n \in J \setminus J_0} a_n \right|.$$

The expression inside the absolute value bars on the right side is essentially a tail of our series, with a finite number of terms missing. Thus, we can certainly say,

$$\left| \sum_{n=N+1}^{\infty} a_n - \sum_{n \in J \setminus J_0} a_n \right| \leq \sum_{n=N+1}^{\infty} |a_n| - \sum_{n \in J \setminus J_0} |a_n| \leq \sum_{n=N+1}^{\infty} |a_n| < \epsilon,$$

as desired. (The differences of summations have the effect of removing certain terms from the infinite series.)

On the other hand, suppose that $\sum_{n=0}^{\infty} |a_n|$ diverges. Suppose for the sake of contradiction that $\sum_{n \in \mathbf{N}} a_n$ converges to a number S, and let J_0 be a finite set such that for every finite set J satisfying $J_0 \subset J \subset I$, we have

$$\left| \sum_{i \in J} a_i - S \right| < 1.$$

Now, since there are only finitely many terms in J_0, $\sum |a_n|$ diverges even if we exclude the terms with indices in J_0. Then we can find a finite collection of terms, indexed by a set K, disjoint from J_0, whose sum is arbitrarily large in absolute value. Then K is a set such that

$$\left| \sum_{n \in K} a_n \right| > 2.$$

This is a contradiction, because,

$$2 < \left| \sum_{n \in K} a_n \right| = \left| \sum_{n \in J_0 \cup K} a_n - \sum_{n \in J_0} a_n \right| \leq \left| \sum_{i \in J_0 \cup K} a_i - S \right| + \left| \sum_{i \in J_0} a_i - S \right| < 1 + 1.$$

Gem 8 *Let $\{a_n\}$ be a sequence of real numbers, and suppose $\sum_{n \in \mathbf{N}} a_n$ diverges properly. Then $\sum_{n=0}^{\infty} a_n$ diverges properly.*

Proof. We appeal to the definitions. Let $N > 0$ be given. Since $\sum_{n \in \mathbf{N}} a_n$ diverges properly, there is a finite set $J_0 \subset \mathbf{N}$ with the property that, for every finite set J satisfying $J_0 \subset$

$J \subset \mathbf{N}$, we have

$$\sum_{i \in J} a_i > N.$$

Then, if we choose M larger than the largest element of J_0, then in particular this guarantees

$$\sum_{n=1}^{M} a_n > N.$$

By definition, this guarantees that $\sum_{n=0}^{\infty} a_n = \infty$.

Gem 9 *The converse to Gem 8 is false. That is, there are properly divergent series $\sum_{n=0}^{\infty} a_n = \infty$ for which $\sum_{n \in \mathbf{N}} a_n$ does not diverge properly.*

Proof. Consider the case $a_n = 1$ for n odd, $a_n = -1/2$ for n even.

Gem 10 *Let I be an uncountable set, and let $\{a_i\}_{i \in I}$ be a sequence indexed by I, and suppose that $\sum_{i \in I} a_i$ converges. Then $a_i = 0$ for all but at most countably many i.*

Proof. Suppose that $\sum_{i \in I} a_i$ converges, and choose some $\epsilon > 0$. Then, by definition we can find a J_0 such that for every finite set J satisfying $J_0 \subset J \subset I$, we have

$$\left| \sum_{i \in J} a_i - S \right| < \epsilon/2.$$

If in particular, $J = J_0 \cup \{i_0\}$ for some $i_0 \notin J_0$ we have

$$\epsilon > \left| \sum_{i \in J} a_i - S \right| + \left| \sum_{i \in J_0} a_i - S \right| \geq \left| \sum_{i \in J} a_i - \sum_{i \in J_0} a_i \right| = |a_{i_0}|.$$

That is, $|a_i| < \epsilon$ for all $i \notin J_0$; in other words there can be only finitely many i such that $|a_i| > \epsilon$ (because all such i must be in J_0). In other words

$$I_\epsilon = \{i \in I \mid |a_i| > \epsilon\}$$

is a finite set. But now we notice

$$\{i \in I \mid a_i \neq 0\} = \bigcup_{n=1}^{\infty} I_{1/n},$$

and so $a_i = 0$ for all but a countable union of finite sets, which is hence countable.

Gem 11 *Let $\{S_i\}_{i \in I}$ be a collection of pairwise disjoint sets, and let $S = \bigcup_{i \in I} S_i$. Let $\{a_s\}_{s \in S}$ be a sequence of positive terms indexed by S. Then*

$$\sum_{i \in I} \sum_{s \in S_i} a_s = \sum_{s \in S} a_s,$$

where we include the possibility that both sides are infinite.

Proof. Just observe that both sides of the equality are equal to

$$\sup_{J \subset S} \sum_{s \in J} a_s,$$

where the supremum is taken over all finite subsets J of S.

Gem 12 *Let $\{S_i\}_{i \in I}$ be a collection of pairwise disjoint sets, and let $S = \bigcup_{i \in I} S_i$. Suppose that $\sum_{s \in S} a_s$ converges as an unordered series. Then for each i, $\sum_{s \in S_i} a_s$ converges as well, and furthermore*

$$\sum_{i \in I} \sum_{s \in S_i} a_s = \sum_{s \in S} a_s.$$

Proof. Based on Gem 10, we can assume that S is countable, so without loss of generality suppose that $S = \mathbf{N}$. By Gem 7, we know $\sum_{n=0}^{\infty} |a_n| < \infty$, so the sum will still be bounded if we restrict the summation to any of the S_i or any finite collection of S_i. Hence $\sum_{n=0; n \in S_i}^{\infty} |a_n| < \infty$, so $\sum_{s \in S_i} a_s$ converges by applying Gem 7 in the opposite direction, and so also $\sum_{i \in I} \sum_{s \in S_i} a_s$ for the same reason.

To see that the sums are equal, we appeal to a trick used in Chapter 1, namely writing $a_s = (a_s + |a_s|) - |a_s|$ as a difference of positive terms. Then our result follows by applying Gem 11 twice, obtaining

$$\sum_{i \in I} \sum_{s \in S_i} (a_s + |a_s|) = \sum_{s \in S} (a_s + |a_s|)$$

and

$$\sum_{i \in I} \sum_{s \in S_i} |a_s| = a \sum_{s \in S} |a_s|.$$

Subtracting gives the desired result.

We remark that these last two results are very powerful, and are largely the reason why unordered series are worth considering. The reordering, regrouping, and rearranging of terms that is covered by this gem is a dramatic generalization of our result that absolutely convergent series can be rearranged without affecting their sum.

Gem 13 *Let $\sum_{n=1}^{\infty} a_n$ be a convergent series or a properly divergent series. Then every grouping of this series converges to the same sum (or is also properly divergent).*

Proof. Recall that, to investigate an infinite series, we investigate its sequence of partial sums. The sequence of partial sums of a grouping of a series is just some subsequence of the partial sums of the original series, and every subsequence of a convergent or properly divergent sequence has the same behavior.

Gem 14 *Let $\sum_{n=1}^{\infty} a_n$ be a series, and consider some grouping of this series that is convergent or properly divergent. Suppose that the sum of the absolute values of the terms in the groups tends to 0. Then the original series converges to the same sum (or is also properly divergent).*

Proof. Suppose that the terms of $\sum_{n=1}^{\infty} a_n$ are grouped as follows,

$$\sum_{i=1}^{\infty}\left(\sum_{n=n_{i-1}+1}^{n_i} a_n\right)$$

for some sequence of integers $0 = n_0 < n_1 < n_2 < \cdots$. Then by hypothesis,

$$\lim_{i\to\infty}\left(\sum_{n=n_{i-1}+1}^{n_i} |a_n|\right) = 0$$

and

$$\sum_{i=1}^{\infty}\left(\sum_{n=n_{i-1}+1}^{n_i} a_n\right)$$

exists. Now, for any N, we associate the integer I so that $n_I < N \leq n_{I+1}$. Then we can estimate partial sums of the original series as follows:

$$\begin{aligned}\left|\sum_{n=1}^{N} a_n - \sum_{i=1}^{I}\left(\sum_{n=n_{i-1}+1}^{n_i} a_n\right)\right| &= \left|\sum_{n=n_I+1}^{N} a_n\right| \\ &\leq \sum_{n=n_I+1}^{N} |a_n| \\ &\leq \sum_{n=n_I+1}^{n_{I+1}} |a_n|\end{aligned}$$

As N goes to infinity, so does I, so we can pass to the limit. The right-hand side is going to 0, so we conclude.

$$\sum_{n=1}^{\infty} a_n = \sum_{i=1}^{\infty}\left(\sum_{n=n_{i-1}+1}^{n_i} a_n\right).$$

Gem 15 *Let $\sum_{n=0}^{\infty} a_n$ be a series, and consider some grouping of this series that is convergent or properly divergent. Suppose that the terms a_n approach 0, and suppose that the* *number of terms in each group is bounded. Then the original series converges to the same sum (or is also properly divergent).*

Proof. This is a corollary of Gem 14. In our grouping, say we never have more than N terms in a group, and choose any $\epsilon > 0$. Then eventually the terms are less than ϵ/N in absolute value, and from that point on the sum of absolute values for any group is less than $N\epsilon/N = \epsilon$. Apply the previous gem.

Gem 16 *Let $\{a_n\}$ be a sequence of numbers greater than 1. Then the infinite series*

$$\sum_{n=1}^{\infty}(a_n - 1)$$

converges if and only if the infinite product

$$\prod_{n=1}^{\infty} a_n$$

converges.

Proof. It is not hard to see that this is equivalent to our Theorem 3.20.

Gem 17 *Let $\{a_n\}$ be a monotone decreasing sequence of positive real numbers and suppose that the series $\sum_{n=1}^{\infty} a_n$ converges. Then we have $\lim_{n\to\infty} na_n = 0$.*

Proof. This result is a strengthening of Theorem 1.44, which would guarantee merely that $\lim_{n\to\infty} a_n = 0$. Since the series is convergent, it must be Cauchy. That is, for each $\epsilon > 0$, there exists an N such that for all $n > m > N$, $\sum_{k=m}^{n} a_k < \epsilon$. Then for each $n > N$, we have

$$\epsilon > \sum_{k=n+1}^{2n} a_k \geq \sum_{k=n+1}^{2n} a_{2n} = na_{2n}$$

so that $2na_{2n} < 2\epsilon$. Also $(2n+1)a_{2n+1} = 2na_{2n+1} + a_{2n+1} \leq 2\epsilon + \epsilon = 3\epsilon$, where the second term of the inequality follows from considering the term a_{2n+1} as the sum $\sum_{k=2n+1}^{2n+1} a_k$. Since we can take $\epsilon \to 0$, we see that for both even and odd n we can make na_n arbitrarily small, as desired.

Gem 18 *The assumption of monotonicity is essential in Gem 17. That is, there are convergent series of positive numbers such that $\lim_{n\to\infty} na_n \neq 0$.*

Proof. Let S be the set $\{1, 4, 9, 16, \dots\}$ of square numbers. We take the example

$$a_n = \begin{cases} \dfrac{1}{n} & \text{when } n \in S \\[2ex] \dfrac{1}{n^2} & \text{otherwise} \end{cases}.$$

Clearly

$$\sum_{n\in S} a_n = \sum_{n\in S} \frac{1}{n} = \sum_{n=1}^{\infty} \frac{1}{n^2}$$

converges. On the other hand,

$$\sum_{n\notin S} a_n = \sum_{n\notin S} \frac{1}{n^2} \leq \sum_{n=1}^{\infty} \frac{1}{n^2}$$

is a convergent p-series. Combining these we see that the series $\sum_{n=1}^{\infty} a_n$ cannot diverge to infinity. Since it has only positive terms it must converge. On the other hand, we notice that $na_n = 1$ for infinitely many n, so that $na_n \not\to 0$.

Gem 19 *The plausible converse to Gem 17 is false. That is, there exists series $\sum_{n=1}^{\infty} a_n$ of positive terms monotonically decreasing to 0 that diverge even though $na_n \to 0$.*

Proof. We simply take $a_n = 1/(n \log n)$. This is defined only for $n > 1$, but we are as always concerned only with the end behavior. We can define $a_1 = 1$ if we please. By the Abel-Dini Scale, this is a divergent series. However,

$$na_n = \frac{1}{\log n},$$

and this clearly decreases monotonically to 0.

Gem 20 *Let $\sum_{n=1}^{\infty} a_n$ and $\sum_{n=1}^{\infty} b_n$ be absolutely convergent series. $\sum_{n=1}^{\infty} a_n b_n$ converges absolutely.*

Proof. By Theorem 1.44, the terms of both series approach 0. Then for sufficiently large n, $|a_n| < 1$. Without loss of generality we can assume that this holds for all n. Then $|a_n b_n| < |b_n|$. The result then follows from the absolute convergence of $\sum_{n=1}^{\infty} b_n$ by the Comparison Test.

Gem 21 (CMJ, 16:2, p. 79) *The series*

$$\sum_{n=1}^{\infty} \left(\frac{1}{2} \frac{3}{4} \frac{5}{6} \cdots \frac{2n-1}{2n} \right)^p$$

converges if and only if $p > 2$.

Proof. We first point out that the Ratio Test would be the natural first choice because each term is defined by adding a new factor to the previous term, so that the ratio between consecutive terms is particularly easily to compute. However, for any p, the Ratio Test would give a limiting ratio of 1 and would therefore be inconclusive. Instead we do something a bit more clever. We introduce a second, similar series that will help us to estimate a_n. We let

$$a_n = \left(\frac{1}{2}\frac{3}{4}\frac{5}{6}\cdots\frac{2n-1}{2n}\right)^p$$

and

$$b_n = \left(\frac{2}{3}\frac{4}{5}\frac{6}{7}\cdots\frac{2n-2}{2n-1}\right)^p.$$

We make some observations. First, in the product $a_n b_n$, most factors appear in numerators and denominators, so that we are left with $a_n b_n = (2n)^{-p}$. Also, we have

$$\frac{1}{2} < \frac{2}{3} < \frac{3}{4} < \frac{4}{5} < \cdots < \frac{2n-3}{2n-2} < \frac{2n-2}{2n-1}, \quad \text{and} \quad \frac{2n-1}{2n} < 1.$$

We combine these inequalities under multiplication to obtain

$$\left(\frac{1}{2}\frac{3}{4}\frac{5}{6}\cdots\frac{2n-1}{2n}\right) < \left(\frac{2}{3}\frac{4}{5}\frac{6}{7}\cdots\frac{2n-2}{2n-1}\right)$$

and

$$\left(\frac{1}{2}\frac{2}{3}\frac{4}{5}\frac{6}{7}\cdots\frac{2n-2}{2n-1}\right) < \left(\frac{1}{2}\frac{3}{4}\frac{5}{6}\cdots\frac{2n-1}{2n}\right),$$

so that, on raising to the p power, $b_n/2 < a_n < b_n$. We now multiply by a_n to obtain

$$\frac{1}{2}(2n)^{-p} = \frac{a_n b_n}{2} < a_n^2 < a_n b_n = (2n)^{-p},$$

or more simply

$$\frac{1}{\sqrt{2}}(2n)^{-p/2} < a_n < (2n)^{-p/2}.$$

Observe that the terms on the left and right side of this inequality are almost identical, differing only by a constant factor. We consider the three quantities being compared as the terms in infinite series, we then know that the left and right series would have the same behavior. By the Comparison Test then, the series $\sum_{n=1}^{\infty} a_n$ would have the same behavior also. In fact the outside series are both, up to a constant factor, p-series with exponent $p/2$. By Theorem 1.59, they converge if and only if $p/2 > 1 \Leftrightarrow p > 2$. We conclude that the given series converges under the same conditions.

Gem 22 (CMJ, 18:5, p. 410) *For any $a > 0, b > 1$, the series*

$$\sum_{n=1}^{\infty} \frac{(\log n)^a}{n^b}$$

converges.

Proof. Intuitively, this result follows because the logarithm function grows much slower than any polynomial, so that even for a very large, the numerator is dominated by n^{ϵ}

for any positive ϵ, so that we should not expect the presence of the numerator to change substantially the conclusion of Theorem 1.59.

To make this explicit, we recall the familiar inequality $\log x \le x - 1$, which holds for all positive x. Taking $x = t^p$, where $t, p > 0$ gives the inequality

$$p \log t = \log(t^p) \le t^p - 1 < t^p,$$

which we rewrite as $\log t < t^p/p$.

We then bound the terms of our series as follows,

$$\frac{(\log n)^a}{n^b} < \frac{(n^p/p)^a}{n^b} = \frac{1}{p^a n^{b-pa}}.$$

We desire to show that the series with these terms converges. The p^a is just a constant multiplier and can be ignored. We see that we have effectively compared our series to a p-series with exponent $b - pa$. However, observe that in invoking the inequality we assumed only that $p > 0$. Then, since $b > 1$, we can choose p sufficiently small that $b - pa > 1$. Implicitly doing so, we can invoke the Comparison Test and we are done.

The following two gems are particularly famous series. As p-series, it is easy to show that they converge, for example by the Integral Test, but to evaluate their sum is quite another matter. Furthermore, notice the unexpected appearance of π in evaluating these sums. Unfortunately, proving in detail these evaluations is a bit beyond the spirit of this chapter, but there are a great variety of proofs in the literature, some quite short and some quite clever. We provide the following list.

In the AMM: Vol. 60, pp. 19–25; Vol. 80, pp. 424–425; Vol. 80, pp. 425–431; Vol. 94, pp. 662–663; Vol. 109, pp. 196–200; Vol. 110, pp. 540–541; Vol. 111, pp. 430–431. In the MM: Vol. 44, pp. 273–276; Vol. 45, pp. 148–149; Vol. 47, pp. 197–202; Vol. 72, pp. 317–319; Vol. 73, pp. 154–155.

Gem 23

$$\sum_{n=1}^{\infty} \frac{1}{n^2} = \frac{\pi^2}{6}.$$

Gem 24

$$\sum_{n=1}^{\infty} \frac{1}{n^4} = \frac{\pi^4}{90}.$$

Gem 25 (CMJ, 17:1, p. 98) *Let S be the set of square numbers $\{1, 4, 9, \dots\}$. Then*

$$\sum_{n \notin S} \frac{1}{n^2} = \frac{\pi^2(15 - \pi^2)}{90}.$$

Proof. Since all the series involved are convergent positive series, it is justified to divide the series as follows.

$$\sum_{n=1}^{\infty} \frac{1}{n^2} = \sum_{n \notin S} \frac{1}{n^2} + \sum_{n \in S} \frac{1}{n^2}.$$

Rearranging,

$$\begin{aligned}
\sum_{n \notin S} \frac{1}{n^2} &= \sum_{n=1}^{\infty} \frac{1}{n^2} - \sum_{n \in S} \frac{1}{n^2} \\
&= \sum_{n=1}^{\infty} \frac{1}{n^2} - \sum_{n=1}^{\infty} \frac{1}{n^4} \\
&= \frac{\pi^2}{6} - \frac{\pi^4}{90} \\
&= \frac{15\pi^2 - \pi^4}{90} \\
&= \frac{\pi^2(15 - \pi^2)}{90}
\end{aligned}$$

as claimed.

Gem 26 (CMJ, 17:1, pp. 98–99) *Let T be the set of square-free positive integers—that is, the set of positive integers not divisible by any square larger than* 1. *Then*

$$\sum_{n \in T} \frac{1}{n^2} = \frac{15}{\pi^2}.$$

Proof. We observe that every positive integer n has a unique factorization $n = m^2 k$, where k is square-free. It follows that

$$\sum_{n=1}^{\infty} \frac{1}{n^2} = \sum_{m=1}^{\infty} \sum_{k \in T} \frac{1}{m^4 k^2}.$$

Since the series is known to be a convergent positive series, we can factor this in the natural way to get

$$\sum_{n=1}^{\infty} \frac{1}{n^2} = \sum_{m=1}^{\infty} \frac{1}{m^4} \sum_{k \in T} \frac{1}{k^2}.$$

The sums over n and m are familiar and we have

$$\frac{\pi^2}{6} = \frac{\pi^4}{90} \sum_{k \in T} \frac{1}{k^2}.$$

The claim follows.

Gem 27

$$1+\frac{1}{5^2}+\frac{1}{7^2}+\frac{1}{11^2}+\cdots=\frac{\pi^2}{9},$$

where the series sums the terms $1/n^2$ *over integers* n *that are not divisible by* 2 *or* 3.

Proof. We will use the Principle of Inclusion-Exclusion. On the intuitive level, our series is formed by beginning with the series $\sum_{n=1}^{\infty} 1/n^2$ and removing the even terms and the multiples of 3. Multiples of 6 have now been eliminated twice, so we add them back once. More explicitly, let U be the set of nonmultiples of 2 and 3. Then we have

$$\begin{aligned}\sum_{n\in U;n\le 6k}\frac{1}{n^2}&=\sum_{n=1}^{6k}\frac{1}{n^2}-\sum_{n=1}^{3k}\frac{1}{(2n)^2}-\sum_{n=1}^{2k}\frac{1}{(3n)^2}+\sum_{n=1}^{k}\frac{1}{(6n)^2}\\&=\sum_{n=1}^{6k}\frac{1}{n^2}-\frac{1}{4}\sum_{n=1}^{3k}\frac{1}{n^2}-\frac{1}{9}\sum_{n=1}^{2k}\frac{1}{n^2}+\frac{1}{36}\sum_{n=1}^{k}\frac{1}{n^2}.\end{aligned}$$

Since the series in question is positive, the partial sums form an increasing sequence. Our analysis technically applies only to the partial sums of our series that stop at or before a multiple of 6, but it is clear that this subcollection of the partial sums will have the same limit behavior as the whole collection. We can therefore allow $k\to\infty$ and, accordingly,

$$\begin{aligned}\sum_{n\in U}\frac{1}{n^2}&=\sum_{n=1}^{\infty}\frac{1}{n^2}-\frac{1}{4}\sum_{n=1}^{\infty}\frac{1}{n^2}-\frac{1}{9}\sum_{n=1}^{\infty}\frac{1}{n^2}+\frac{1}{36}\sum_{n=1}^{\infty}\frac{1}{n^2}\\&=\left(1-\frac{1}{4}-\frac{1}{9}+\frac{1}{36}\right)\sum_{n=1}^{\infty}\frac{1}{n^2}\\&=\frac{2}{3}\cdot\frac{\pi^2}{6}\\&=\frac{\pi^2}{9},\end{aligned}$$

where we have invoked Gem 23. Notice that this type of argument could easily be applied to a problem in which multiples of other combinations of numbers were removed.

Gem 28 *Suppose that* $\sum_{n=1}^{\infty} a_n$ *converges while* $\sum_{n=1}^{\infty} a_n^2$ *diverges. Then* $\sum_{n=1}^{\infty} a_n$ *converges conditionally.*

Proof. Assume for the sake of contradiction that $\sum_{n=1}^{\infty} a_n$ converges absolutely, i.e., that $\sum_{n=1}^{\infty} |a_n|$ converges. By Theorem 1.44, we must have $a_n\to 0$, so that eventually we have $|a_n|<1$ and so $a_n^2<|a_n|$. By the Comparison Test, we then have $\sum_{n=1}^{\infty} a_n^2$ as a convergent series, a contradiction.

Gem 29

$$\sum_{n=1}^{\infty} \frac{(-1)^{n+1}}{n^n} = \int_0^1 x^x \, dx.$$

Proof. We require the power series for e^x, as well as the fact that we can integrate power series term by term.

$$\begin{aligned}
\int_0^1 x^x \, dx &= \int_0^1 e^{x \log x} \, dx \\
&= \int_0^1 \sum_{n=0}^{\infty} \frac{(x \log x)^n}{n!} \, dx \\
&= \sum_{n=0}^{\infty} \int_0^1 \frac{(x \log x)^n}{n!} \, dx.
\end{aligned}$$

Writing $S_{j,k}(x) = x^j (\log x)^k$, we proceed to compute $\int_0^1 S_{j,k}(x) \, dx$. We claim that

$$\int_0^1 S_{j,k}(x) \, dx = \frac{(-1)^k k!}{(j+1)^{k+1}},$$

which we prove by induction on k. The case $k = 0$ is clear by the power rule. For the inductive step, we have, by integration by parts,

$$\begin{aligned}
\int_0^1 S_{j,k}(x) \, dx &= \frac{1}{j+1} S_{j+1,k} \Big]_0^1 - \frac{k}{j+1} \int_0^1 S_{j,k-1}(x) \, dx \\
&= \frac{-k}{j+1} \int_0^1 S_{j,k-1}(x) \, dx \\
&= \frac{-k}{j+1} \frac{(-1)^{k-1}(k-1)!}{(j+1)^k} \\
&= \frac{(-1)^k k!}{(j+1)^{k+1}}.
\end{aligned}$$

We are using here the easily proven fact that $\lim_{x \to 0^+} S_{j,k}(x) = 0$ for $j > 0$ and all k.

We now finish the calculations as follows:

$$\int_0^1 x^x \, dx = \sum_{n=0}^{\infty} \frac{1}{n!} \int_0^1 S_{n,n}(x) \, dx = \sum_{n=0}^{\infty} \frac{1}{n!} \frac{(-1)^n n!}{(n+1)^{n+1}} = \sum_{n=1}^{\infty} \frac{(-1)^{n+1}}{n^n}.$$

Gem 30 (AMM, 59:2, pp. 108–109)

$$\int_0^1 \frac{dx}{x^x} = \sum_{i=1}^{\infty} \frac{1}{i^i}.$$

Proof. We follow the line of Gem 29 and use freely the results proven there.

$$\begin{aligned}\int_0^1 \frac{dx}{x^x} &= \int_0^1 e^{-x\log x}\,dx \\ &= \int_0^1 \sum_{n=0}^{\infty} \frac{(-x\log x)^n}{n!}\,dx \\ &= \sum_{n=0}^{\infty} \int_0^1 \frac{(-x\log x)^n}{n!}\,dx.\end{aligned}$$

We now finish the calculations as follows:

$$\int_0^1 \frac{dx}{x^x} = \sum_{n=0}^{\infty} \frac{(-1)^n}{n!} \int_0^1 S_{n,n}(x)\,dx = \sum_{n=0}^{\infty} \frac{(-1)^n}{n!}\frac{(-1)^n n!}{(n+1)^{n+1}} = \sum_{n=1}^{\infty} \frac{1}{n^n}.$$

Gem 31

$$\sum_{n=1}^{\infty} \frac{1}{n^{1+1/n}}$$

diverges.

Proof. We first note that $2^n > n$. The cases $n = 0, 1$ are easy to check. For $n \geq 2$, we write n as the product of n factors

$$n = 1 \cdot 1 \cdot \frac{2}{1} \cdot \frac{3}{2} \cdot \cdots \cdot \frac{n}{n-1}$$

and notice that each of the factors is less than 2 (except $2/1$, which equals 2), and we are done. (A more clever proof is to recognize this as a special case of the very well-known set theory result that the power set of a set is always strictly larger than the original set.)

Now we can bound our series from below using the Comparison Test.

$$\sum_{n=1}^{\infty} \frac{1}{n^{1+1/n}} = \sum_{n=1}^{\infty} \frac{1}{n \cdot n^{1/n}} \geq \sum_{n=1}^{\infty} \frac{1}{2n}.$$

This last series, though, is just half the divergent harmonic series. The given series is even larger, so it too must diverge.

Gem 32 *A series of the form*

$$\sum_{n=1}^{\infty} \frac{1}{n^{1+a_n}}$$

with $a_n > 0$ *and* $\lim_{n\to\infty} a_n = 0$ *could converge or diverge.*

Proof. This result is included to counter a common notion among calculus students. It is natural (but wrong) to view the series in the statement of the result as a p-series with exponent greater than 1. However, the statement and proof of that theorem relied on p being a constant. Combining this test with the Comparison Test, we conclude that a series of the form described above would certainly converge if the a_n had a lower bound larger than 0. However, nothing in the statement of the p-series test shows that the series described in the result should diverge. Still we might want to improve our proof technique to include this case. This result shows that no such result is possible because either behavior is possible. A divergent example was given in Gem 31. A convergent result comes by taking $a_n = 2\log(\log n)/\log n$ so that $n^{a_n} = (\log n)^2$, where we have used basic properties of the logarithm. Of course we must restrict our attention now to n sufficiently large that the iterated logarithm is defined. We leave to the reader to show that these numbers are positive and tend to 0. However,

$$\sum_{n=3}^{\infty} \frac{1}{n^{1+a_n}} = \sum_{n=3}^{\infty} \frac{1}{n(n^{a_n})}$$
$$= \sum_{n=3}^{\infty} \frac{1}{n(\log n)^2}$$

which converges by the Abel-Dini Scale.

Gem 33

$$\sum_{n=1}^{\infty} \operatorname{Arctan}\left(\frac{1}{2n^2}\right) = \frac{\pi}{4}.$$

Proof. We recall the trigonometric identity

$$\tan(\alpha - \beta) = \frac{\tan\alpha - \tan\beta}{1 + \tan\alpha \tan\beta}$$

or equivalently

$$\operatorname{Arctan} x - \operatorname{Arctan} y = \operatorname{Arctan}\left(\frac{x-y}{1+xy}\right)$$

whenever $-\pi < \operatorname{Arctan} x - \operatorname{Arctan} y < \pi$. In particular

$$\operatorname{Arctan}\frac{1}{2n-1} - \operatorname{Arctan}\frac{1}{2n+1} = \operatorname{Arctan}\left(\frac{\frac{1}{2n-1} - \frac{1}{2n+1}}{1 + \frac{1}{2n-1}\frac{1}{2n+1}}\right)$$
$$= \operatorname{Arctan}\left(\frac{(2n+1)-(2n-1)}{(2n+1)(2n-1)+1}\right)$$
$$= \operatorname{Arctan}\left(\frac{2}{4n^2}\right)$$
$$= \operatorname{Arctan}\left(\frac{1}{2n^2}\right).$$

We therefore write our series in the form

$$\sum_{n=1}^{\infty}\left(\operatorname{Arctan}\frac{1}{2n-1}-\operatorname{Arctan}\frac{1}{2n+1}\right)$$

and notice that it telescopes. That is, writing S_n for the nth partial sum, we have

$$S_n = \operatorname{Arctan} 1 - \operatorname{Arctan}\frac{1}{2n+1}.$$

Since $\tan 0 = 0$, the right term vanishes in the limit, and

$$\sum_{n=1}^{\infty}\operatorname{Arctan}\left(\frac{1}{2n^2}\right) = \lim_{n\to\infty} S_n = \operatorname{Arctan} 1 = \frac{\pi}{4}.$$

Gem 34 (CMJ, 21:3, pp. 253–254)

$$\sum_{n=1}^{\infty}\operatorname{Arctan}\left(\frac{2}{n^2}\right) = \frac{3\pi}{4}.$$

Proof. As in Gem 33, we have

$$\operatorname{Arctan} x - \operatorname{Arctan} y = \operatorname{Arctan}\left(\frac{x-y}{1+xy}\right).$$

That is,

$$\begin{aligned}\operatorname{Arctan}\frac{1}{n-1} - \operatorname{Arctan}\frac{1}{n+1} &= \operatorname{Arctan}\left(\frac{\frac{1}{n-1}-\frac{1}{n+1}}{1+\frac{1}{n-1}\frac{1}{n+1}}\right)\\ &= \operatorname{Arctan}\left(\frac{(n+1)-(2n-1)}{(n+1)(n-1)+1}\right)\\ &= \operatorname{Arctan}\left(\frac{2}{n^2}\right).\end{aligned}$$

We therefore write our series in the form

$$\operatorname{Arctan} 2 + \sum_{n=2}^{\infty}\left(\operatorname{Arctan}\frac{1}{n-1} - \operatorname{Arctan}\frac{1}{n+1}\right)$$

and notice that it telescopes. That is, writing S_n for the nth partial sum, we have

$$S_n = \operatorname{Arctan} 2 + \operatorname{Arctan} 1 + \operatorname{Arctan}\frac{1}{2} - \operatorname{Arctan}\frac{1}{n} - \operatorname{Arctan}\frac{1}{n+1}.$$

Since $\tan 0 = 0$, the right terms vanish in the limit, and

$$\begin{aligned}\sum_{n=1}^{\infty} \operatorname{Arctan}\left(\frac{2}{n^2}\right) &= \lim_{n\to\infty} S_n \\ &= \operatorname{Arctan} 2 + \operatorname{Arctan} 1 + \operatorname{Arctan}\frac{1}{2} \\ &= \operatorname{Arctan} 2 + \operatorname{Arctan}\frac{1}{2} + \frac{\pi}{4}.\end{aligned}$$

However, the inverse tangents of 2 and $1/2$ are complementary angles! So we have

$$\sum_{n=1}^{\infty} \operatorname{Arctan}\left(\frac{2}{n^2}\right) = \frac{\pi}{2} + \frac{\pi}{4} = \frac{3\pi}{4}.$$

Gem 35

$$\sum_{k=0}^{\infty} \frac{k^2+3k+1}{(k+2)!} = 2.$$

Proof. We just notice that

$$\frac{k^2+3k+1}{(k+2)!} = \frac{(k+1)(k+2)-1}{(k+2)!} = \frac{1}{k!} - \frac{1}{(k+2)!}$$

so that our series telescopes,

$$\sum_{k=0}^{n} \frac{k^2+3k+1}{(k+2)!} = \frac{1}{0!} + \frac{1}{1!} - \frac{1}{(n+1)!} - \frac{1}{(n+2)!}.$$

Passing to the limit,

$$\sum_{k=0}^{\infty} \frac{k^2+3k+1}{(k+2)!} = \frac{1}{0!} + \frac{1}{1!} = 2$$

as desired.

Gem 36 (MM, 32:4, p. 229)

$$\sum_{n=1}^{\infty} 6^{(2-3n-n^2)/2}$$

converges to an irrational number.

Proof. Convergence is evident from comparison with the convergent geometric series $\sum_{n=0}^{\infty} 6^{-n}$. To investigate irrationality, we operate in base 6. Then each term of the series contributes a 1 to the heximal expansion of the sum. We notice that the exponents in the terms are $-1, -4, -8, -13, \ldots$. The difference between consecutive terms is constantly changing by -1. That is, the sum will be $0.1001000100001\ldots$ where each successive block of zeros has one more zero. Suppose to the contrary that the sum is rational, so that this expansion is eventually repeating. However, if the eventual block has d digits, the block will eventually be entirely contained in a block of z zeros, where $z > 2d - 1$. Thus the digits of the expansion must be eventually constantly 0. But this is clearly false. Thus this series has an irrational sum. (Note the similarity to Liouville's irrational number, which is $0.101001000100001000001\ldots$, in base 10.)

Definition 4.2 *A power series centered at a is a series of functions of the form*

$$\sum_{n=0}^{\infty} c_n (x-a)^n$$

for some constants $\{c_n\}$. *For the purposes of this definition,* $(x-a)^0$ *is always taken to be* 1, *even in the case* $x = a$. *For our purposes, we will consider only the case when* $a = 0$.

We wish to use some basic facts about power series, not because this book intends to deal directly with series of functions, but because we can use power series as a vehicle to learn about series of constants. Because it is series of real constants that we wish to focus on here, we will not provide proofs for the following propositions or for Gems 37 through 41. The interested reader can consult any mainstream beginning calculus or advanced calculus textbook. We particularly recommend the excellent books by James Stewart and Witold Kosmala (included in our references).

Proposition 4.3 *Let*

$$\sum_{n=0}^{\infty} c_n x^n$$

be a power series. Then exactly one of the following occurs.

1. *The series converges only for* $x = 0$.
2. *There is a positive number R such that the series converges for* $|x| < R$ *and diverges for* $|x| > R$.
3. *The series converges for all x.*

Proposition 4.4 *A power series converges absolutely wherever it converges, except possibly at the endpoints of the interval of convergence.*

Definition 4.5 *If a power series* $\sum_{n=0}^{\infty} c_n x^n$ *is convergent for* $|x| < r$, $r > 0$, *and moreover*

$$f(x) = \sum_{n=0}^{\infty} c_n x^n$$

for some function f and all $|x| < r$, *we say that the power series represents the function f.*

Proposition 4.6 *A function cannot be represented by distinct power series with the same center. That is, if*

$$\sum_{n=0}^{\infty} a_n x_n = \sum_{n=0}^{\infty} b_n x^n$$

on some interval $|x| < r$, then $a_n = b_n$ for all n.

Proposition 4.7 *If a function f is represented by a power series centered at 0 that converges for $|x| < r$, then f has continuous derivatives of all orders at 0, and*

$$f(x) = \sum_{n=0}^{\infty} f^{(n)}(0) \frac{x^n}{n!},$$

where $f^{(n)}$ denotes the nth derivative.

Proposition 4.8 *Power series can be added and subtracted term by term. Power series can be multiplied using the Cauchy formula,*

$$\left(\sum_{n=0}^{\infty} a_n x^n\right)\left(\sum_{n=0}^{\infty} b_n x^n\right) = \sum_{n=0}^{\infty}\left(\sum_{k=0}^{n} a_k b_{n-k}\right) x^n.$$

Proposition 4.9 *Power series can be differentiated term by term. If*

$$f(x) = \sum_{n=0}^{\infty} a_n x^n$$

for $|x| < R$, then

$$f'(x) = \sum_{n=0}^{\infty} n a_n x^{n-1} \quad \text{for } |x| < R.$$

Proposition 4.10 *Power series can be integrated term by term. If*

$$f(x) = \sum_{n=0}^{\infty} a_n x^n$$

for $|x| < R$, then

$$\int_0^x f(t)dt = \sum_{n=0}^{\infty} a_n \frac{x^{n+1}}{n+1} \quad \text{for } |x| < R.$$

Gem 37

$$\sum_{n=0}^{\infty} \frac{x^n}{n!} = e^x \quad \text{for all } x.$$

Gem 38

$$\sum_{n=0}^{\infty} \frac{(-1)^n x^{2n+1}}{(2n+1)!} = \sin x \quad \textit{for all } x.$$

Gem 39

$$\sum_{n=0}^{\infty} \frac{(-1)^n x^{2n}}{(2n)!} = \cos x \quad \textit{for all } x.$$

Gem 40

$$\sum_{n=1}^{\infty} \frac{(-1)^{n+1} x^n}{n} = \log(1+x) \quad \textit{for all } -1 < x \leq 1.$$

Gem 41

$$\sum_{n=0}^{\infty} x^n = \frac{1}{1-x} \quad \textit{for all } -1 < x < 1.$$

Gem 42 *For any q satisfying $|q| > 1$, we have*

$$\sum_{n=1}^{\infty} \frac{n}{q^n} = \frac{q}{(q-1)^2}.$$

Proof. Beginning with the result in Gem 41, we differentiate both sides to obtain

$$\sum_{n=1}^{\infty} nx^{n-1} = \frac{1}{(1-x)^2}.$$

Multiplying by x, we have

$$\sum_{n=1}^{\infty} nx^n = \frac{x}{(1-x)^2}.$$

This relation holds for all $|x| < 1$. In particular we can take $x = 1/q$, and obtain

$$\sum_{n=1}^{\infty} \frac{n}{q^n} = \frac{1/q}{(1-1/q)^2} = \frac{q}{(q-1)^2}.$$

Gem 43 *Let p be any polynomial function, and let c be any constant satisfying $|c| < 1$. Then the series*

$$\sum_{n=1}^{\infty} p(n)c^n$$

converges. Moreover, there is an algorithm for computing the sum in closed form. Furthermore, if p has rational coefficients and c is rational, the sum will also be rational.

Proof. Consider the polynomials

$$p_0(n) = 1, \qquad p_k(n) = n(n-1)(n-2)\cdots(n-k+1).$$

We claim that any polynomial

$$p(n) = a_m n^m + a_{m-1}n^{m-1} + \cdots + a_1 n + a_0$$

can be expressed as a sum

$$p(n) = b_m p_m(n) + b_{m-1}p_{m-1}(n) + \cdots + b_1 p_1(n) + b_0 p_0(n),$$

and furthermore if the coefficients a_i are rational, so also are the b_i.

This is easily proven by induction on m. The base case $m = 0$ is trivial. Just take $b_0 = a_0$. Assume that the proposition holds for all $m < N$. Consider any polynomial

$$p(n) = a_N n^N + a_{N-1}n^{N-1} + \cdots + a_1 n + a_0.$$

It is clear that this polynomial has leading term $a_N n^N$, it is also clear that, if we expand $a_N p_N(n)$, its leading term will be the same. Further, the coefficients of $a_N p_N(n)$ are rational if a_N is. Then we can write

$$p(n) = a_N p_N(n) + q(n)$$

for some polynomial $q(n)$ of degree at most $N-1$, whose coefficients are rational if the a_i are. By induction, $q(n)$ can be expressed as a sum of the $p_m(n)$ in the desired way, and we are done with this claim. Notice that this proof gives the algorithm for computing the b_i. Take $b_m = a_m$, expand $b_m p_m(n)$, subtract to get a new polynomial; let b_{m-1} be the new leading coefficient, and so on.

Then every series of the form

$$\sum_{n=1}^{\infty} p(n)c^n$$

will be a linear combination of series of the form

$$\sum_{n=1}^{\infty} p_m(n)c^n,$$

and it will suffice to prove that all of these sums converge, and that the sum is rational if c is.

Beginning with Gem 41, differentiate m times to pass from

$$\frac{1}{1-x} = \sum_{n=0}^{\infty} \frac{x^n}{n!}$$

to

$$\frac{m!}{(1-x)^{m+1}} = \sum_{n=m}^{\infty} n(n-1)(n-2)\cdots(n-m+1)x^{n-m} = \sum_{n=m}^{\infty} p_m(n)x^{n-m}.$$

Multiply by x^m to obtain

$$\frac{m!x^m}{(1-x)^{m+1}} = \sum_{n=m}^{\infty} p_m(n)x^n = \sum_{n=0}^{\infty} p_m(n)x^n.$$

This holds for all x with $|x| < 1$, so we can in particular take $x = c$, to obtain

$$\sum_{n=1}^{\infty} p_m(n)c^n = \frac{m!c^m}{(1-c)^{m+1}}.$$

Since we know all the values when $p = p_m$, and since any polynomial can be decomposed into a sum of p_i by a simple algorithm, the result is established.

Gem 44

$$\sum_{n=1}^{\infty} \frac{2^n n^2}{n!} = 6e^2.$$

Proof. By Gem 37, we have

$$e^x = \sum_{n=0}^{\infty} \frac{x^n}{n!} = \sum_{n=1}^{\infty} \frac{x^{n-1}}{(n-1)!}.$$

Multiply by x to get

$$xe^x = \sum_{n=1}^{\infty} \frac{x^n}{(n-1)!} = \sum_{n=1}^{\infty} \frac{nx^n}{n!}.$$

Differentiate term-by-term, and then multiply by x for the following.

$$xe^x + e^x = \sum_{n=1}^{\infty} \frac{n^2 x^{n-1}}{n!},$$

$$x^2e^x + xe^x = \sum_{n=1}^{\infty} \frac{n^2 x^n}{n!}.$$

The above equality holds for all real x, so we can specialize to $x = 2$ to conclude that

$$\sum_{n=1}^{\infty} \frac{2^n n^2}{n!} = 6e^2.$$

Gem 45 *Let p be any polynomial function, and let c be any constant satisfying $|c| < 1$. Then the series*

$$\sum_{n=1}^{\infty} p(n)\frac{c^n}{n!}$$

converges. Moreover, there is an algorithm for computing the sum in closed form. Furthermore, if p has rational coefficients and c is rational, the sum will be a rational multiple of e^c.

Proof. Recall the polynomials $p_0(n) = 1$, $p_k(n) = n(n-1)(n-2)\cdots(n-k+1)$ from the proof of Gem 43. Using the work done in that proof, every series of the form

$$\sum_{n=1}^{\infty} p(n)\frac{c^n}{n!}$$

will be a linear combination of series of the form

$$\sum_{n=1}^{\infty} p_m(n)\frac{c^n}{n!},$$

and it will suffice to prove that all of these sums converge, and that the sum is a rational multiple of e^c when c is rational.

Beginning with Gem 37, differentiate m times to pass from

$$e^x = \sum_{n=0}^{\infty} \frac{x^n}{n!}$$

to

$$e^x = \sum_{n=m}^{\infty} n(n-1)(n-2)\cdots(n-m+1)\frac{x^{n-m}}{n!} = \sum_{n=m}^{\infty} p_m(n)\frac{x^{n-m}}{n!}.$$

Multiply by x^m to obtain

$$x^m e^x = \sum_{n=m}^{\infty} p_m(n)\frac{x^n}{n!} = \sum_{n=0}^{\infty} p_m(n)\frac{x^n}{n!}.$$

This holds for all x, so we can in particular take $x = c$, to obtain

$$\sum_{n=1}^{\infty} p_m(n)\frac{c^n}{n!} = c^m e^c.$$

Since we know all the values when $p = p_m$, and since any polynomial can be decomposed into a sum of p_i by a simple algorithm, we have done all we claimed.

Gem 46 *Consider the series*

$$\frac{1}{2} + \frac{1}{3} + \frac{1}{5} + \frac{1}{7} + \frac{1}{11} + \cdots$$

where we add reciprocals of primes. Then no partial sum can be an integer.

Proof. In fact we prove the stronger result that no finite collection of terms from this series can have an integer sum. Let $p_1, p_2, \ldots, p_n$ be a collection of distinct primes, and let $N = p_1 p_2 \cdots p_{n-1}$, where we say $N = 1$ if $n = 1$. Assume for the sake of contradiction that

$$\frac{1}{p_1} + \frac{1}{p_2} + \frac{1}{p_3} + \cdots + \frac{1}{p_n}$$

is an integer. Then it will remain an integer after multiplying by the integer N to get

$$\frac{N}{p_1} + \frac{N}{p_2} + \frac{N}{p_3} + \cdots + \frac{N}{p_n}.$$

However, in this expression all the terms are integers except the last one, which cannot be. Then the sum is not an integer.

Gem 47 (CMJ, 28:4, pp. 296–297) *Let $\sum_{n=1}^{\infty} a_n$ be a convergent series of positive terms. Then there exists another convergent series $\sum_{n=1}^{\infty} A_n$ that decreases more slowly in the sense that*

$$\lim_{n\to\infty} \frac{A_n}{a_n} = \infty.$$

Proof. On an intuitive level, the Comparison Test guarantees that series of positive terms will converge if their terms go to 0 sufficiently quickly. That is, if two series $\sum_{n=1}^{\infty} a_n$ and $\sum_{n=1}^{\infty} b_n$ satisfy $a_n/b_n \to \infty$, then the terms b_n go to 0 faster, and the convergence of $\sum_{n=1}^{\infty} a_n$ implies the convergence of $\sum_{n=1}^{\infty} b_n$. We might desire a "slowest" convergent series $\sum_{n=1}^{\infty} a_n$, with the property that any convergent positive series would then be shown to converge by the Ratio Comparison Test. This result shows in effect that this is unrealizable, because any convergent series is dominated by another series that also converges. Here we use "dominated" in a strong sense, namely that the limit of the ratio is infinite.

For the construction, we write t_n for the nth "tail" of the series, i.e., $t_n = a_n + a_{n+1} + a_{n+2} + \cdots$, which all exist because the series converges. Then we notice that the sequence $\{t_n\}$ is a sequence of positive numbers decreasing to 0. Now we define $A_n = a_n/\sqrt{t_n}$. We

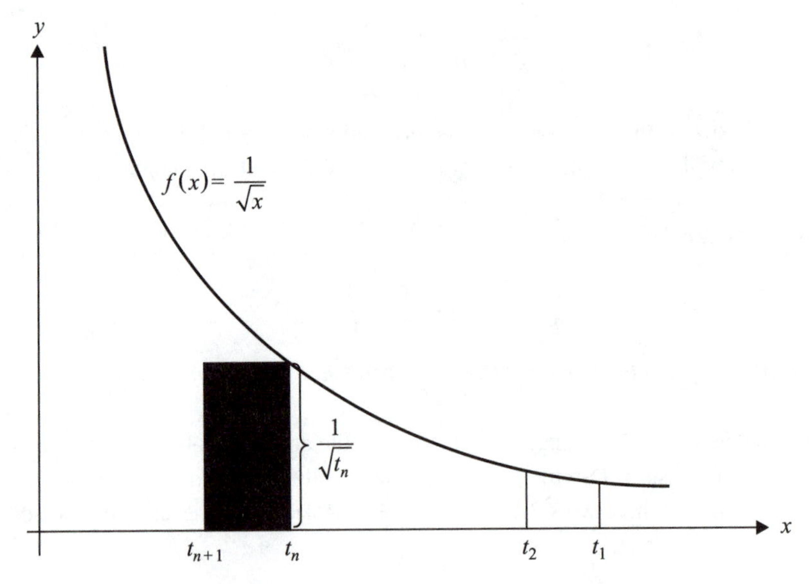

FIGURE 4.1
Viewing the Terms A_n as Rectangles in a Riemann Sum

claim that these A_n satisfy the claim. First

$$\lim_{n\to\infty}\frac{A_n}{a_n}=\lim_{n\to\infty}\frac{1}{\sqrt{t_n}}=\infty,$$

so that we need only show that $\sum_{n=1}^{\infty} A_n$ converges. The key is to interpret the sum as a Riemann sum. Since

$$A_n=\frac{a_n}{\sqrt{t_n}}=\frac{t_n-t_{n+1}}{\sqrt{t_n}},$$

each term in the series can be viewed as the area of a rectangle in Figure 4.1. As shown in the figure, each partial sum of $\sum_{n=1}^{\infty} A_n$ is bounded above by the area

$$\int_0^{t_1}\frac{1}{\sqrt{x}}\,dx=2\sqrt{t_1}.$$

The series is therefore positive with bounded partial sums, hence convergent.

Gem 48 (CMJ, 28:4, pp. 296–297) *Let $\sum_{n=1}^{\infty} D_n$ be a divergent series of positive terms. Then there exists another divergent series $\sum_{n=1}^{\infty} d_n$ that is smaller in the sense that*

$$\lim_{n\to\infty}\frac{D_n}{d_n}=\infty.$$

Proof. This is of course the dual result to Gem 47. Here we show that there is no smallest divergent series. We use a dual construction to the one used there. Let s_n be the nth partial sum of $\sum_{n=1}^{\infty} D_n$, and define $d_n = D_n/s_{n-1}$. We claim that this definition of d_n satisfies our claim. (This is not defined for $n = 1$. Here we will not concern ourselves with that. The reader who insists on remedying this may define d_1 any way he or she pleases.) First

$$\lim_{n\to\infty}\frac{D_n}{d_n}=\lim_{n\to\infty}s_{n-1}=\infty,$$

so that we need only show that $\sum_{n=2}^{\infty} d_n$ diverges. The key is to interpret the sum as a Riemann sum. Since

$$d_n=\frac{D_n}{s_{n-1}}=\frac{s_n-s_{n-1}}{s_{n-1}},$$

each term in the series can be viewed as the area of a rectangle in Figure 4.2. As shown in the figure, each partial sum of $\sum_{n=2}^{\infty} d_n$ is bounded below by the area

$$\int_{s_1}^{s_n}\frac{1}{x}\,dx=\log s_n-\log s_1.$$

As $n\to\infty$, we see that this lower bound also increases without bound, so that this series diverges.

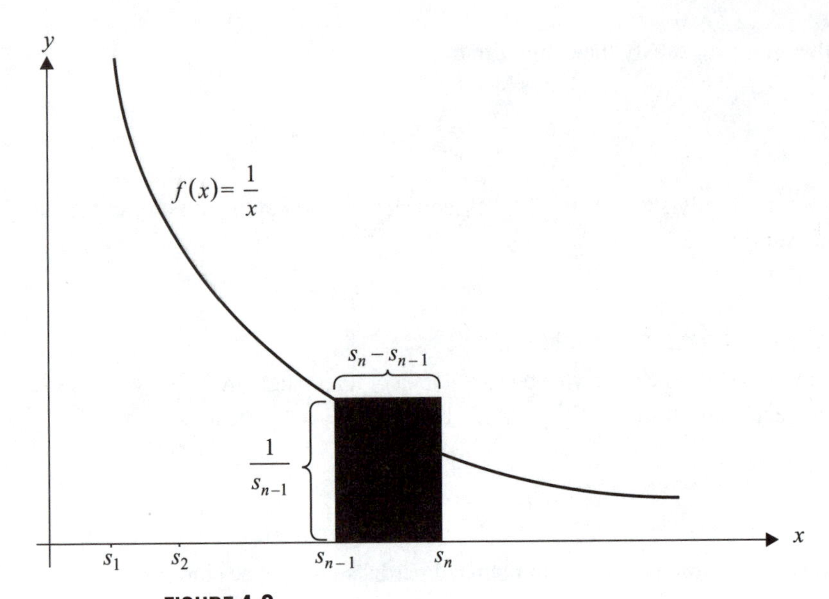

FIGURE 4.2
Viewing the Terms d_n as Rectangles in a Riemann Sum

Gem 49 *Define a sequence by* $a_1 = 2$, $a_{n+1} = a_n^2 - a_n + 1$. *Then*

$$\sum_{n=1}^{\infty} \frac{1}{a_n} = 1.$$

Proof. We notice that $a_{n+1} - 1 = a_n(a_n - 1)$, so that

$$\frac{1}{a_{n+1} - 1} = \frac{1}{a_n(a_n - 1)} = \frac{1}{a_n - 1} - \frac{1}{a_n}$$

or equivalently

$$\frac{1}{a_n} = \frac{1}{a_n - 1} - \frac{1}{a_{n+1} - 1}.$$

Noting that this telescopes, we have

$$S_n = \frac{1}{a_1 - 1} - \frac{1}{a_{n+1} - 1} = 1 - \frac{1}{a_{n+1} - 1}.$$

It now suffices to show that the sequence $\{a_n\}$ increases without bound. Since it is a sequence of integers, it will suffice to show that it is increasing. However, $a_{n+1} - a_n = a_n^2 - 2a_n + 1 = (a_n - 1)^2 > 0$, and we are done.

Gem 50

$$\sum_{n=1}^{\infty} \frac{1}{(n!)^2}$$

converges to an irrational number.

Proof. First, this series clearly converges by comparison with the convergent series $\sum_{n=1}^{\infty} 1/n^2$. Now suppose for the sake of contradiction that this series converges to a rational number p/q. Then

$$\begin{aligned} q[(q-1)!]^2 p &= (p/q)(q!)^2 \\ &= (q!)^2 \sum_{n=1}^{\infty} \frac{1}{(n!)^2} \\ &= \sum_{n=1}^{\infty} \left(\frac{q!}{n!}\right)^2 \\ &= \sum_{n=1}^{q} \left(\frac{q!}{n!}\right)^2 + \sum_{n=q+1}^{\infty} \left(\frac{q!}{n!}\right)^2 . \end{aligned}$$

Since the left-hand side and every term in the left summation are integers, we must also have the right summation be an integer. But

$$\begin{aligned} 0 &< \sum_{n=q+1}^{\infty} \left(\frac{q!}{n!}\right)^2 \\ &= \frac{1}{(q+1)^2} + \frac{1}{(q+1)^2(q+2)^2} + \frac{1}{(q+1)^2(q+2)^2(q+3)^2} + \cdots \\ &< \frac{1}{(q+1)^2}\left(1 + \frac{1}{(q+2)^2} + \frac{1}{(q+2)^4} + \cdots\right) \\ &= \frac{1}{(q+1)^2}\left(\frac{1}{1-1/(q+2)^2}\right) \\ &= \frac{(q+2)^2}{(q+1)^2[(q+2)^2-1]}. \end{aligned}$$

Since the denominator is quartic and the numerator is quadratic, this last fraction is eventually arbitrarily small; in particular, for q sufficiently large, it is too small to be a positive integer. However, by writing our rational number p/q in an unreduced form if necessary, q can be chosen arbitrarily large. Thus for that q,

$$0 < \sum_{n=q+1}^{\infty} \left(\frac{q!}{n!}\right)^2 < 1,$$

a contradiction.

Gem 51 (MM, 17:7 pp. 292–295)

$$\sum_{n=1}^{\infty} \frac{1}{1^2 + 2^2 + 3^2 + \cdots + n^2} = 18 - 24 \log 2.$$

Proof. We first recall that

$$1^2 + 2^2 + 3^2 + \cdots + n^2 = \frac{n(n+1)(2n+1)}{6},$$

which can be proven in a straightforward way using mathematical induction. Thus we can write

$$\sum_{n=1}^{\infty} \frac{1}{1^2 + 2^2 + 3^2 + \cdots + n^2} = \sum_{n=1}^{\infty} \frac{6}{n(n+1)(2n+1)}$$
$$= \sum_{n=1}^{\infty} \left(\frac{6}{n} + \frac{6}{n+1} - \frac{24}{2n+1} \right).$$

Writing S_n for the nth partial sum of the series, we can perform some manipulations on the partial sums as follows

$$S_n = \left(\frac{6}{1} + \frac{6}{2} - \frac{24}{3} \right) + \left(\frac{6}{2} + \frac{6}{3} - \frac{24}{5} \right) + \left(\frac{6}{3} + \frac{6}{4} - \frac{24}{7} \right)$$
$$+ \cdots + \left(\frac{6}{n} + \frac{6}{n+1} - \frac{24}{2n+1} \right)$$
$$= 6 - \frac{24}{3} + \left(\frac{6}{2} + \frac{6}{2} \right) - \frac{24}{5} + \left(\frac{6}{3} + \frac{6}{3} \right) - \frac{24}{7} + \left(\frac{6}{4} + \frac{6}{4} \right)$$
$$+ \cdots + \left(\frac{6}{n} + \frac{6}{n} \right) - \frac{24}{2n+1} + \frac{6}{n+1}$$
$$= 6 - 24 \left(\frac{1}{3} - \frac{1}{4} + \frac{1}{5} - \frac{1}{6} + \frac{1}{7} - \cdots + \frac{1}{2n+1} \right) + \frac{6}{n+1}$$

We recognize the similarity between the expression in parentheses and the partial sums of the alternating harmonic series, which motivates us to write

$$S_n = 6 + \frac{24}{1} - \frac{24}{2} - 24 \left(\frac{1}{1} - \frac{1}{2} + \frac{1}{3} - \frac{1}{4} + \frac{1}{5} - \cdots + \frac{1}{2n+1} \right) + \frac{6}{n+1}.$$

Referencing Theorem 3.6, we can pass to the limit and write

$$\sum_{n=1}^{\infty} \frac{1}{1^2 + 2^2 + 3^2 + \cdots + n^2} = 18 - 24 \log 2.$$

Gem 52

$$\sum_{n=1}^{\infty} \frac{1}{n^2 + n/2} = 4(1 - \log 2).$$

Proof.

$$\begin{aligned}\sum_{n=1}^{k} \frac{1}{n^2+n/2} &= \sum_{n=1}^{k} \frac{4}{(2n)(2n+1)} \\ &= 4\sum_{n=1}^{k}\left(\frac{1}{2n}-\frac{1}{2n+1}\right) \\ &= 4\left(1-\sum_{n=1}^{2k+1}\frac{(-1)^{n+1}}{n}\right).\end{aligned}$$

We have thus reduced summing this series to summing the alternating harmonic series, which we have already effected by Theorem 3.6. Thus

$$\sum_{n=1}^{\infty} \frac{1}{n^2+n/2} = 4(1-\log 2)$$

as desired.

Gem 53 *Let $\sum_{n=1}^{\infty} a_n$ be a convergent positive series, and define $t_k = \sum_{n=k}^{\infty} a_n$ to be the kth tail of the series. Suppose further that $\sum_{n=1}^{\infty} t_n$ converges. Then*

$$\lim_{n\to\infty} n a_n = 0.$$

Proof. Since the terms are positive, the tails form a monotone decreasing sequence (each tail includes one fewer positive term than the previous tail). Applying Gem 17, we have

$$\lim_{n\to\infty} n t_n = 0.$$

We simply notice $0 < n a_n < n t_n$ and apply the Squeeze Principle.

(Note: An alternative proof of this result can be derived by combining Gem 54 and the Corollary 1.44.1.)

Gem 54 *Let $\sum_{n=1}^{\infty} a_n$ be a convergent positive series, and define $t_k = \sum_{n=k}^{\infty} a_n$ to be the kth tail of the series. Then*

$$\sum_{n=1}^{\infty} t_n = \sum_{n=1}^{\infty} n a_n,$$

where we include the possibility where both sides are infinite together.

Proof. We write

$$\sum_{n=1}^{\infty} t_n = \sum_{n=1}^{\infty}\sum_{k=n}^{\infty} a_k$$

and

$$\sum_{n=1}^{\infty} na_n = \sum_{k=1}^{\infty}\sum_{n=1}^{k} a_k.$$

Now we notice that both of the above summations contain the same terms (a_k occurs in each summation for each combination $n \le k$). Since all the terms are positive, the only possibilities are proper divergence or absolute convergence. In the former case, both summations will be infinite. In the latter case, the above rearrangements will not affect the sum.

Gem 55 (CMJ, 17:2, p. 188) *Let H_n be the nth partial sum of the harmonic series, and let n_p be the smallest integer n satisfying $H_n \ge p$. Then*

$$\lim_{p\to\infty} \frac{n_{p+1}}{n_p} = e.$$

Proof. Intuitively this is plausible. By Theorem 3.3 we know that H_n grows like $\log n$, so that we would expect n_p, which is in a sense an "approximate inverse" to H_n, to grow like e^p. We justify this as follows. Since

$$H_{n_p} - \frac{1}{n_p} = H_{n_p-1} < p \le H_{n_p},$$

we have $H_{n_p} = p - k_p$, where k_p is a nonnegative constant bounded above by $1/n_p$. In particular $k_p \to 0$. Recalling the notation of Theorem 3.3, we have

$$p = H_{n_p} + k_p = \log n_p + \frac{1}{n_p} + \gamma_{n_p} + k_p$$

from which it follows that

$$\lim_{p\to\infty} (p - \log n_p) = \lim_{p\to\infty} \left(\frac{1}{n_p} + \gamma_{n_p} + k_p\right) = \gamma.$$

Similarly we have also

$$\lim_{p\to\infty} (p + 1 - \log n_{p+1}) = \gamma.$$

Subtracting, we have

$$\lim_{p\to\infty} [(p + 1 - \log n_{p+1}) - (p - \log n_p)] = 0.$$

Rearranging,

$$\lim_{p\to\infty} \left(1 - \log \frac{n_{p+1}}{n_p}\right) = 0,$$

which is equivalent to the proposed

$$\lim_{p\to\infty} \frac{n_{p+1}}{n_p} = e.$$

Gem 56 *Let $\sum_{n=1}^{\infty} a_n$ be a convergent series of positive terms. Then*

$$\sum_{n=1}^{\infty} \sqrt{a_n a_{n+1}}$$

converges also.

Proof. By the familiar result that the arithmetic mean bounds above the geometric mean, we have

$$\sqrt{a_n a_{n+1}} \leq \frac{1}{2}(a_n + a_{n+1}).$$

But $\sum_{n=1}^{\infty} a_n$ and so also $\sum_{n=1}^{\infty} a_{n+1}$ converges. Thus $\sum_{n=1}^{\infty}(a_n + a_{n+1})/2$ converges, so that the Comparison Test applies.

Gem 57 *Suppose that $\{a_n\}$ is a monotone decreasing sequence of positive terms, and suppose that*

$$\sum_{n=1}^{\infty} \sqrt{a_n a_{n+1}}$$

converges. Then $\sum_{n=1}^{\infty} a_n$ converges also.

Proof. This is a weakened converse to Gem 56. We simply notice that $a_{n+1} \leq \sqrt{a_n a_{n+1}}$ and apply the Comparison Test.

Gem 58 *Suppose that $\{a_n\}$ is a sequence of positive terms, and suppose that*

$$\sum_{n=1}^{\infty} \sqrt{a_n a_{n+1}}$$

converges. Then $\sum_{n=1}^{\infty} a_n$ does not necessarily converge.

Proof. This result demonstrates that the potential converse to Gem 56 is actually false. Consider

$$a_n = \begin{cases} 1 & \text{if } n \text{ is odd} \\ \dfrac{1}{n^4} & \text{if } n \text{ is even} \end{cases}.$$

Since the series

$$\sum_{n=1}^{\infty} \sqrt{a_n a_{n+1}}$$

is positive, we can simply check that its partial sums are bounded. We have

$$\sum_{n=1}^{k} \sqrt{a_n a_{n+1}} < \sum_{n=1}^{2k} \sqrt{a_n a_{n+1}} = \sum_{n=1}^{k} \left(\frac{1}{(2n)^2} + \frac{1}{(2n)^2} \right) = \frac{1}{2} \sum_{n=1}^{k} \frac{1}{n^2}.$$

The right-hand side, however, is bounded because the summations are the partial sums of a convergent p-series. However, $\sum_{n=1}^{\infty} a_n$ diverges because its terms do not go to 0.

Gem 59 *In Gems 56, 57, and 58, we can replace* $\sqrt{a_n a_{n+1}}$ *by the expression*

$$\sqrt[p]{a_n a_{n+1} a_{n+2} \cdots a_{n+p-1}}$$

for each $p = 3, 4, \ldots$. *The analogues of these results continue to hold under these replacements.*

Proof. The first two results hold with a virtually identical proof to that presented above. For the third, we proceed as above, using the following counterexample.

$$a_n = \begin{cases} 1 & \text{if } n \text{ is not a multiple of } p \\ \dfrac{1}{n^{2p}} & \text{if } n \text{ is a multiple of } p \end{cases}.$$

Gem 60

$$\sum_{n=1}^{\infty} \sin \frac{1}{2^{n+1}} \cos \frac{3}{2^{n+1}} = \frac{1}{2} \sin 1.$$

Proof. First, we recall a familiar result from trigonometry, the product-to-sum identity

$$\sin \alpha \cos \beta = \frac{1}{2} [\sin(\beta + \alpha) - \sin(\beta - \alpha)].$$

Using this identity we can rewrite our series in a telescoping form.

$$\sum_{n=1}^{\infty} \sin \frac{1}{2^{n+1}} \cos \frac{3}{2^{n+1}} = \sum_{n=1}^{\infty} \left(\sin \frac{1}{2^{n-1}} - \sin \frac{1}{2^n} \right).$$

Partial sums, then, have the form

$$\frac{1}{2} \left(\sin 1 - \sin \frac{1}{2^n} \right),$$

which tends to the claimed $(1/2) \sin 1$.

Gem 61

$$\sum_{n=1}^{\infty} \frac{n}{1 \cdot 3 \cdot 5 \cdot 7 \cdots (2n+1)} = \frac{1}{2}.$$

Proof. Writing

$$a_n = \frac{n}{1 \cdot 3 \cdot 5 \cdot 7 \cdots (2n+1)}$$

for the terms of our series, we can rewrite

$$\begin{aligned} a_n &= \frac{1}{2}\left(\frac{(2n+1)-1}{1 \cdot 3 \cdot 5 \cdot 7 \cdots (2n+1)}\right) \\ &= \frac{1}{2}\left(\frac{2n+1}{1 \cdot 3 \cdot 5 \cdot 7 \cdots (2n+1)} - \frac{1}{1 \cdot 3 \cdot 5 \cdot 7 \cdots (2n+1)}\right) \\ &= \frac{1}{2}\left(\frac{1}{1 \cdot 3 \cdot 5 \cdot 7 \cdots (2n-1)} - \frac{1}{1 \cdot 3 \cdot 5 \cdot 7 \cdots (2n+1)}\right). \end{aligned}$$

Once again this series exhibits a telescoping behavior, so that our partial sums have the form

$$\frac{1}{2}\left(\frac{1}{1} - \frac{1}{1 \cdot 3 \cdot 5 \cdot 7 \cdots (2n+1)}\right).$$

As n goes to ∞, this tends to a limit of $1/2$.

Gem 62 (CMJ, 15:5, pp. 448–450) *For any positive integers n and p, we have*

$$\sum_{m=n}^{\infty} \frac{1}{m(m+1)(m+2)\cdots(m+p)} = \frac{1}{pn(n+1)(n+2)\cdots(n+p-1)}.$$

Proof. Once again, we manipulate the terms of the series into a telescoping form. If we write

$$a_m = \frac{1}{m(m+1)(m+2)\cdots(m+p-1)},$$

we have

$$\begin{aligned} a_m - a_{m+1} &= \frac{1}{m(m+1)(m+2)\cdots(m+p-1)} - \frac{1}{(m+1)(m+2)\cdots(m+p)} \\ &= \frac{m+p}{m(m+1)(m+2)\cdots(m+p)} - \frac{m}{m(m+1)(m+2)\cdots(m+p)} \\ &= \frac{p}{m(m+1)(m+2)\cdots(m+p)}. \end{aligned}$$

Using this, we can write

$$\sum_{m=n}^{\infty} \frac{1}{m(m+1)(m+2)\cdots(m+p)} = \sum_{m=n}^{\infty} \frac{1}{p}(a_m - a_{m+1}).$$

The partial sums, then, have the form

$$\sum_{m=n}^{N} \frac{1}{m(m+1)(m+2)\cdots(m+p)} = \frac{1}{p}(a_n - a_{N+1})$$

where we remember that in this case n is a constant. As $N \to \infty$, the second member on the right-hand side vanishes. Thus the sum of the series is

$$\frac{a_n}{p} = \frac{1}{pn(n+1)(n+2)\cdots(n+p-1)}.$$

Remark. We have used the phenomenon of telescoping to evaluate and diagnose a variety of series. In the next gem, we see that even in series that are merely similar to telescoping series, the telescoping can be exploited.

Gem 63

$$\sum_{n=1}^{\infty} \frac{\sin\frac{1}{n} - \sin\frac{1}{n+1}}{\sqrt[3]{n}}$$

converges.

Proof. Although this series is erratic and does not telescope, we notice that the numerators have a telescoping form. Let

$$a_n = \sin\frac{1}{n} - \sin\frac{1}{n+1}.$$

This telescopes, and we have

$$\sum_{n=1}^{\infty} a_n = \sin 1.$$

In particular the series converges. Since

$$a_n \geq \frac{a_n}{\sqrt[3]{n}} = \frac{\sin\frac{1}{n} - \sin\frac{1}{n+1}}{\sqrt[3]{n}},$$

the claim follows from the Comparison Test.

Gem 64 *The series*

$$1 + \frac{1}{2} - \frac{1}{3} + \frac{1}{4} + \frac{1}{5} - \frac{1}{6} + \frac{1}{7} + \frac{1}{8} - \frac{1}{9} + \cdots$$

diverges properly.

Proof. Since the pattern of signs repeats every three terms, it is natural to consider only the partial sums that include entire blocks of three terms. (That is, we consider summations that terminate after a negative term.) We leave it to the reader to confirm that such an analysis is sufficient to understand the behavior of the series.

$$S_{3n} = \sum_{k=1}^{n} \left(\frac{1}{3k-2} + \frac{1}{3k-1} - \frac{1}{3k} \right) = \sum_{k=1}^{n} \left(\frac{9k^2 - 2}{(3k-2)(3k-1)3k} \right).$$

Then our series will have the limit behavior of

$$\sum_{k=1}^{\infty}\left(\frac{9k^2-2}{(3k-2)(3k-1)3k}\right);$$

however, this series diverges to ∞ by the Limit Comparison Test against the harmonic series.

Gem 65 *The series*

$$1+\frac{1}{2}-\frac{2}{3}+\frac{1}{4}+\frac{1}{5}-\frac{2}{6}+\frac{1}{7}+\frac{1}{8}-\frac{2}{9}+\cdots$$

converges to $\log 3$.

Proof. As in Gem 64, we consider only partial sums of the form S_{3n}. In this case we have

$$\begin{aligned}
S_{3n} &= 1+\frac{1}{2}-\frac{2}{3}+\frac{1}{4}+\frac{1}{5}-\frac{2}{6}+\cdots+\frac{1}{3n-2}+\frac{1}{3n-1}-\frac{2}{3n}\\
&= 1+\frac{1}{2}+\frac{1}{3}+\frac{1}{4}+\frac{1}{5}+\frac{1}{6}+\cdots+\frac{1}{3n}-\frac{3}{3}-\frac{3}{6}-\cdots-\frac{3}{3n}\\
&= \sum_{k=1}^{3n}\frac{1}{k}-\sum_{k=1}^{n}\frac{1}{k}.
\end{aligned}$$

Using Theorem 3.3, we have

$$\begin{aligned}
S_{3n} &= \log 3n+\gamma_{3n}+\frac{1}{3n}-\log n-\gamma_n-\frac{1}{n}\\
&= \log 3+\gamma_{3n}-\gamma_n-\frac{2}{3n}.
\end{aligned}$$

In the limit, the γ expressions cancel and we have a sum of $\log 3$.

Gem 66 (MM, 60:2, p. 118) *Let* $H_n=\sum_{k=1}^{n}1/k$. *Then the series*

$$\sum_{k=1}^{\infty}\frac{1}{kH_k}$$

diverges.

Proof. We will show that the series is not Cauchy. In particular, given any positive integer N we show there is an integer m such that

$$\sum_{k=N}^{N+m}\frac{1}{kH_k}>1/2.$$

The key is to notice the near-telescoping behavior of this series.

$$\begin{aligned}\sum_{k=N}^{N+m} \frac{1}{kH_k} &= \sum_{k=N}^{N+m} \frac{H_k - H_{k-1}}{H_k} \\ &\geq \sum_{k=N}^{N+m} \frac{H_k - H_{k-1}}{H_{N+m}} \\ &= \frac{H_{N+m} - H_{N-1}}{H_{N+m}} \\ &= 1 - \frac{H_{N-1}}{H_{N+m}},\end{aligned}$$

which is certainly larger than $1/2$ for m large enough. (The harmonic series diverges, so H_{m+n} goes to infinity in m and is in particular eventually larger than $2H_{N-1}$.

Gem 67 (Red Book, Problem 99)[3] *Let* $H_n = 1 + 1/2 + 1/3 + \cdots + 1/n$. *Then*

$$\sum_{n=1}^{\infty} \frac{H_n}{n(n+1)} = \frac{\pi^2}{6}.$$

Proof. Let k be a positive integer. We have

$$\begin{aligned}\sum_{n=1}^{k} \frac{H_n}{n(n+1)} &= \sum_{n=1}^{k} \left(\frac{H_n}{n} - \frac{H_n}{n+1}\right) \\ &= \sum_{n=1}^{k} \frac{H_n}{n} - \sum_{n=2}^{k+1} \frac{H_{n-1}}{n} \\ &= H_1 + \sum_{n=2}^{k} \frac{H_n - H_{n-1}}{n} - \frac{H_k}{k+1} \\ &= 1 + \sum_{n=2}^{k} \frac{1}{n^2} - \frac{H_k}{k+1} \\ &= \sum_{n=1}^{k} \frac{1}{n^2} - \frac{H_k}{k+1}.\end{aligned}$$

Passing to the limit, the first term in the above expression evaluates to $\pi^2/6$ by Gem 23. The second term goes to 0 because $H_k = \log k + \gamma_k + 1/k$ as in Theorem 3.3, which is essentially logarithmic and hence dominated by $k+1$.

---

[3]We wish to thank Dover Publications for allowing us to use three problems from *The Red Book of Mathematical Problems* and one problem from *The Green Book of Mathematical Problems*, both co-authored by Kenneth Williams and Kenneth Hardy. These appear as our Gems 67, 70, 71 and 72.

Gem 68 (AMM, 59:7, p. 471) *Writing as usual* $H_n = \sum_{k=1}^{n} 1/k$ *and* $H_0 = 0$, *we have*

$$\sum_{n=1}^{\infty} \frac{H_n}{n^2} = 2\sum_{n=1}^{\infty} \frac{1}{n^3} = 2\sum_{n=1}^{\infty} \frac{H_n}{(n+1)^2}.$$

Proof. It will be enough to prove the first and third expressions equal, since then we will be done by

$$\begin{aligned}\sum_{n=1}^{\infty} \frac{H_n}{n^2} &= \sum_{n=1}^{\infty} \frac{H_{n-1} + 1/n}{n^2} \\ &= \sum_{n=1}^{\infty} \frac{H_{n-1}}{n^2} + \sum_{n=1}^{\infty} \frac{1}{n^3} \\ &= \sum_{n=1}^{\infty} \frac{H_n}{(n+1)^2} + \sum_{n=1}^{\infty} \frac{1}{n^3}.\end{aligned}$$

To this end, we have

$$\begin{aligned}\sum_{n=1}^{\infty} \frac{H_n}{n^2} &= \sum_{n=1}^{\infty} \frac{1}{n^2} \sum_{k=1}^{n} \frac{1}{k} \\ &= \sum_{n=1}^{\infty} \frac{1}{n^2} \sum_{k=1}^{\infty} \left(\frac{1}{k} - \frac{1}{n+k}\right) \\ &= \sum_{n=1}^{\infty} \sum_{k=1}^{\infty} \frac{1}{n^2} \left(\frac{1}{k} - \frac{1}{n+k}\right) \\ &= \sum_{n=1}^{\infty} \sum_{k=1}^{\infty} \frac{1}{nk(n+k)} \\ &= \sum_{n=1}^{\infty} \sum_{k=1}^{\infty} \left(\frac{1}{k(n+k)^2} + \frac{1}{n(n+k)^2}\right) \\ &= 2\sum_{n=1}^{\infty} \sum_{k=1}^{\infty} \frac{1}{k(n+k)^2}\end{aligned}$$

by symmetry. We now change variables so that $k = R$ and $k + n = N + 1$. As k and n run over positive integers, N runs over the positive integers, and R runs over the positive integers strictly less than N, so we have

$$2\sum_{N=1}^{\infty} \sum_{R=1}^{N} \frac{1}{R(N+1)^2} = 2\sum_{N=1}^{\infty} \frac{H_N}{(N+1)^2}.$$

Gem 69 (AMM, 59:7, pp. 471–472)

$$\sum_{K,N} \frac{1}{N^2 K} = 2,$$

where the summation is taken over all positive integers K and N with $K \leq N$ and K relatively prime to N.

Proof. Every pair of positive integers (k, n) with $k \leq n$ can be uniquely expressed in the form (rK, rN) where $K \leq N$ and K is relatively prime to N. (Let r be the greatest common divisor of k and n.) Now, allowing ourselves to freely reorder the positive terms of this series, we use Gem 68 to obtain

$$\begin{aligned} 2\left(\sum_{r=1}^{\infty} \frac{1}{r^3}\right) &= \sum_{n=1}^{\infty} \frac{H_n}{n^2} \\ &= \sum_{n=1}^{\infty} \sum_{k=1}^{n} \frac{1}{kn^2} \\ &= \sum_{r,K,N} \frac{1}{(rK)(rN)^2} \\ &= \sum_{r,K,N} \frac{1}{r^3} \frac{1}{KN^2} \\ &= \left(\sum_{r=1}^{\infty} \frac{1}{r^3}\right)\left(\sum_{K,N} \frac{1}{N^2 K}\right) \end{aligned}$$

and we are done since all the series involved converge to finite positive sums. In executing the final factorization, we are using the observation that all r are possible, and that whether a particular pair K, N is included in the summation does not depend on r.

Gem 70 (Red Book, Problem 18) *Let*

$$a_n = \frac{1}{4n+1} + \frac{1}{4n+3} - \frac{1}{2n+2}.$$

The series $\sum_{n=0}^{\infty} a_n$ converges to $(3/2)\log 2$.

Proof. Let $s_N = \sum_{n=0}^{N} a_n$. Then

$$\begin{aligned} s_N &= \sum_{n=0}^{N} \left(\frac{1}{4n+1} + \frac{1}{4n+3} - \frac{1}{2n+2}\right) \\ &= \sum_{n=0}^{N} \left(\frac{1}{4n+1} - \frac{1}{4n+2} + \frac{1}{4n+3} - \frac{1}{4n+4} + \frac{1}{4n+2} - \frac{1}{4n+4}\right) \\ &= \sum_{m=1}^{4N+4} \frac{(-1)^{m-1}}{m} + \frac{1}{2} \sum_{m=1}^{2N+2} \frac{(-1)^{m-1}}{m}. \end{aligned}$$

Passing to the limit of large N, each of these summations becomes the alternating harmonic series; we have

$$\lim_{N\to\infty} s_N = \frac{3}{2}\sum_{m=1}^{\infty}\frac{(-1)^{m-1}}{m} = \frac{3}{2}\log 2.$$

Gem 71 (Red Book, Problem 62) *Let $a > 1$. Then*

$$\sum_{n=0}^{\infty}\frac{2^n}{a^{2^n}+1} = \frac{1}{a-1}.$$

Proof. We have

$$\begin{aligned}\frac{2^n}{a^{2^n}+1} &= \frac{2^n(a^{2^n}-1)}{a^{2^{n+1}}-1}\\ &= \frac{2^n(a^{2^n}+1)-2^{n+1}}{a^{2^{n+1}}-1}\\ &= \frac{2^n}{a^{2^n}-1} - \frac{2^{n+1}}{a^{2^{n+1}}-1},\end{aligned}$$

so that we have a telescoping series as follows:

$$\sum_{n=0}^{\infty}\frac{2^n}{a^{2^n}+1} = \sum_{n=0}^{\infty}\left(\frac{2^n}{a^{2^n}-1} - \frac{2^{n+1}}{a^{2^{n+1}}-1}\right) = \frac{1}{a-1}.$$

(To verify that this is legitimate, we must check that $2^n/(a^{2^n}-1)$ tends to 0 for large n. But then this is a subsequence of $N/(a^N-1)$ and certainly the denominator, which grows exponentially since $a > 1$, dominates the linear numerator.)

Gem 72 (Green Book, Problem 16)

$$\sum_{n=0}^{\infty}\frac{(-1)^n(n+1)^3}{n!} = \frac{1}{1} - \frac{8}{1} + \frac{27}{2} - \frac{64}{6} + \cdots = -e^{-1}.$$

Proof. We begin with the polynomial identity

$$(n+1)^3 = n(n-1)(n-2) + 6n(n-1) + 7n + 1$$

so that

$$\begin{aligned}
\sum_{n=0}^{\infty} \frac{(-1)^n(n+1)^3}{n!} &= \sum_{n=0}^{\infty} (-1)^n \frac{n(n-1)(n-2)+6n(n-1)+7n+1}{n!} \\
&= \sum_{n=3}^{\infty} \frac{(-1)^n}{(n-3)!} + 6\sum_{n=2}^{\infty} \frac{(-1)^n}{(n-2)!} + 7\sum_{n=1}^{\infty} \frac{(-1)^n}{(n-1)!} + \sum_{n=0}^{\infty} \frac{(-1)^n}{n!} \\
&= -\sum_{n=0}^{\infty} \frac{(-1)^n}{n!} + 6\sum_{n=0}^{\infty} \frac{(-1)^n}{n!} - 7\sum_{n=0}^{\infty} \frac{(-1)^n}{n!} + \sum_{n=0}^{\infty} \frac{(-1)^n}{n!} \\
&= -\sum_{n=0}^{\infty} \frac{(-1)^n}{n!} \\
&= -e^{-1}.
\end{aligned}$$

Remark. The next results concern the Cauchy product of series. If $\sum_{n=0}^{\infty} a_n$ and $\sum_{n=0}^{\infty} b_n$ are series, then it is natural to consider the product

$$\left(\sum_{n=0}^{\infty} a_n\right)\left(\sum_{n=0}^{\infty} b_n\right) = \sum_{m,n=0}^{\infty} a_m b_n.$$

However, this is a series indexed by pairs of natural numbers rather than by natural numbers. Since evaluating a series depends on the term order, this infinite summation is not obviously well-defined. Of course if all the terms are positive or more generally if $\sum_{m,n=0}^{\infty} a_m b_n$ converges as an *unordered* series, we can avoid this issue. We would like, though, to distinguish one particular way of assigning meaning to $\sum_{m,n=0}^{\infty} a_m b_n$ even when convergence is not absolute. The following definition is motivated by considering the manner in which like terms are grouped when multiplying polynomials and power series.

Definition 4.11 *The Cauchy product of the series $\sum_{n=0}^{\infty} a_n$ and $\sum_{n=0}^{\infty} b_n$ is given by*

$$\left(\sum_{n=0}^{\infty} a_n\right)\left(\sum_{n=0}^{\infty} b_n\right) = \sum_{n=0}^{\infty}\left(\sum_{k=0}^{n} a_k b_{n-k}\right) = \sum_{n=0}^{\infty} c_n,$$

where $c_n = \sum_{k=0}^{n} a_k b_{n-k}$.

Gem 73 *The Cauchy product of two convergent series may be divergent.*

Proof. We take as our example $a_n = b_n = (-1)^{n+1}/\sqrt{n+1}$. By the Alternating Series Test, both of these series converge. Let $c_n = \sum_{k=0}^{n} a_k b_{n-k}$, so that $\sum_{n=0}^{\infty} c_n$ is the Cauchy product in question, and indeed

$$c_n = \sum_{n=0}^{\infty} \frac{(-1)^{k+1}}{\sqrt{k+1}} \frac{(-1)^{n-k+1}}{\sqrt{n-k+1}} = (-1)^n \sum_{k=0}^{n} \frac{1}{\sqrt{k+1}\sqrt{n-k+1}}.$$

Hence

$$|c_n| = \sum_{k=0}^{n} \frac{1}{\sqrt{k+1}\sqrt{n-k+1}} \geq \sum_{k=0}^{n} \frac{1}{\sqrt{n+1}\sqrt{n+1}} = \sum_{k=0}^{n} \frac{1}{n+1} = 1,$$

so that the terms of the Cauchy product do not approach 0. The Cauchy series, it is now clear, diverges by oscillation.

Gem 74 *The Cauchy product of two divergent series may converge absolutely.*

Proof. Let $a_0 = 3$ and $a_n = 3^n$ for positive n. Let $b_0 = -2$ and $b_n = 2^n$ for positive n. Let $c_n = \sum_{k=0}^{n} a_k b_{n-k}$, so that $\sum_{n=0}^{\infty} c_n$ is the Cauchy product in question. Then of course $c_0 = -6$, and for positive n,

$$\begin{aligned} c_n &= \sum_{k=0}^{n} a_k b_{n-k} \\ &= 3 \cdot 2^n + \sum_{k=1}^{n-1} 2^k 3^{n-k} - 2 \cdot 3^n \\ &= 3 \cdot 2^n + 3^n \sum_{k=1}^{n-1} (2/3)^k - 2 \cdot 3^n \\ &= 3 \cdot 2^n + 3^n [2/3 - (2/3)^n]/(1 - 2/3) - 2 \cdot 3^n \\ &= 3 \cdot 2^n + 2 \cdot 3^n - 3 \cdot 2^n - 2 \cdot 3^n \\ &= 0, \end{aligned}$$

so that the terms of the Cauchy product are 0 after the first, and so the product is absolutely convergent.

Gem 75 *The Cauchy product of an absolutely convergent series and a convergent series will converge. Furthermore, its sum will be the ordinary product of the sums of the original series.*

Proof. This is a classical result due to Mertens, and we do not prove it here. See, for example, T. M. Apostol's *Mathematical Analysis*, 2nd ed., pp. 204–205. (A full citation is given in the appendices.)

Gem 76 *The Cauchy product is associative.*

Proof. Let $\sum_{n=0}^{\infty} a_n$, $\sum_{n=0}^{\infty} b_n$, $\sum_{n=0}^{\infty} c_n$ be three series, and define new terms as follows.

$$d_n = \sum_{k=0}^{n} a_k b_{n-k},$$

$$e_n = \sum_{k=0}^{n} b_k c_{n-k}.$$

(That is, in the Cauchy sense $\sum d_n = \left(\sum a_n\right)\left(\sum b_n\right)$ and $\sum e_n = \left(\sum b_n\right)\left(\sum c_n\right)$.)

$$f_n = \sum_{k=0}^{n} a_k e_{n-k},$$

$$g_n = \sum_{k=0}^{n} d_k c_{n-k}.$$

(That is, in the Cauchy sense $\sum f_n = \sum a_n \left(\sum b_n \sum c_n\right)$ and $\sum g_n = \left(\sum a_n \sum b_n\right) \sum c_n$.)

It will suffice now to show that $f_n = g_n$. But it is not hard to check that

$$f_n = \sum_{k=0}^{n} a_k e_{n-k} = \sum_{k=0}^{n} a_k \sum_{j=0}^{n-k} b_j c_{n-k-j} = \sum_{k=0}^{n}\sum_{j=0}^{n-k} a_k b_j c_{n-k-j}$$

and

$$g_n = \sum_{k=0}^{n} d_k b_{n-k} = \sum_{k=0}^{n}\left(\sum_{j=0}^{k} a_j b_{k-j}\right) c_{n-k} = \sum_{k=0}^{n}\sum_{j=0}^{k} a_j b_{k-j} c_{n-k}.$$

But each of these is just the same as $\sum_{j+k+l=n} a_j b_k c_l$.

Gem 77 (CMJ, 22:1, pp. 71–72) *The infinite product*

$$\prod_{n=1}^{\infty} \frac{(n^2+1)}{\sqrt{n^4+4}} = \sqrt{2}.$$

Proof. First, we recall the factorization

$$n^4 + 4 = (n^2+2n+2)(n^2-2n+2) = [(n+1)^2+1][(n-1)^2+1].$$

Then we have

$$\prod_{n=1}^{N} \frac{n^2+1}{\sqrt{n^4+4}} = \prod_{n=1}^{N} \frac{\sqrt{n^2+1}}{\sqrt{[(n+1)^2+1]}} \frac{\sqrt{n^2+1}}{\sqrt{[(n-1)^2+1]}}.$$

We recognize this as a telescoping product.

$$\prod_{n=1}^{N} \frac{n^2+1}{\sqrt{n^4+4}} = \frac{\sqrt{2}\sqrt{N^2+1}}{\sqrt{(N+1)^2+1}}.$$

Passing to the limit, we have

$$\prod_{n=1}^{\infty} \frac{n^2+1}{\sqrt{n^4+4}} = \lim_{N\to\infty} \frac{\sqrt{2}\sqrt{N^2+1}}{\sqrt{(N+1)^2+1}} = \sqrt{2}.$$

Gem 78 *If $\sum_{n=1}^{\infty} a_n$ converges, then $\lim_{n\to\infty}(a_n + a_{n+1} + \cdots + a_{n+p}) = 0$ for every nonnegative integer p.*

Proof. We can regard this result as a strengthened version of Theorem 1.44, and also as a weakened but convenient version of the Cauchy criterion. Since this series converges, the Cauchy criterion applies. Then for every $\epsilon > 0$ there exists a positive integer N such that for all $m \geq n \geq N$ we have $|\sum_{k=n}^{m} a_k| < \epsilon$. In particular this is still true if we additionally stipulate that $m = n + p + 1$. Then eventually $|a_n + a_{n+1} + \cdots + a_{n+p}| < \epsilon$. Thus the limit is 0, as desired.

This "miniature" Cauchy criterion is often useful because the sequence $\{(a_n + a_{n+1} + a_{n+2} + \cdots + a_{n+p})\}$ may be very well-behaved for some particular p even if $\{a_n\}$ is very badly behaved.

Gem 79 *The converse to Gem 78 fails. There exist series $\sum_{n=1}^{\infty} a_n$ that are divergent but such that $\lim_{n\to\infty}(a_n + a_{n+1} + \cdots + a_{n+p}) = 0$ for every nonnegative integer p.*

Proof. Take the harmonic series as a counterexample.

$$0 \leq \lim_{n\to\infty} \left(\frac{1}{n} + \frac{1}{n+1} + \cdots + \frac{1}{n+p}\right) \leq \lim_{n\to\infty} \frac{p+1}{n} = 0$$

as desired, but of course the harmonic series is divergent.

Gem 80 (CMJ, 28:3, p. 232) *Let $f(t) = t - at^2$ for some $a > 0$ Write f^n for the composition of f with itself n times, where by convention f^0 is the identity function, and f^1 is f itself. Then $\sum_{n=0}^{\infty} f^n(t)$ converges if and only if $t = 0$ or $t = 1/a$.*

Proof. First of all, if $t = 0$ or $t = 1/a$, the terms in the series are eventually 0, so that the series converges.

Suppose $t > 1/a$. Then $f(t) = t - at^2 < 0$. But also, for any $x \neq 0$, $f(x) = x - ax^2 < x$. Inductively now we have

$$f^{n+1}(t) < f^n(t) < \cdots < f^2(t) < f(t) < 0.$$

Thus the terms do not tend to 0, guaranteeing divergence. The same argument applies if $t < 0$.

We have only to consider now the case $0 < t < 1/a$. Then $0 < f(t) \leq 1/(4a)$. Then, choose N large enough that $1/(2Na) < f(t)$. Since f is strictly increasing on $(0, \frac{1}{4a}]$ (by observing that the derivative, $1 - 2at > 0$), we know

$$f^2(t) > f\left(\frac{1}{2Na}\right) = \frac{1}{2Na} - \frac{a}{4N^2a^2} = \frac{2N-1}{4N^2a} \geq \frac{1}{(2N+2)a}.$$

By a straightforward induction, for $n \geq 1$ we have $f^n(t) \geq 1/[(2N+2n-2)a]$. The series then diverges by applying, say, the limit comparison test against the harmonic series.

Gem 81 (CMJ, 28:2, pp. 149–150) *Let $p > 1$ and write $\zeta(p)$ for the value of the p-series*

$$\sum_{n=1}^{\infty} \frac{1}{n^p}.$$

Then we can evaluate the following series:

$$\sum_{n=1}^{\infty} \frac{1}{n^p}\left(1+\frac{1}{2^p}+\frac{1}{3^p}+\cdots+\frac{1}{n^p}\right)=\frac{1}{2}[\zeta(p)^2+\zeta(2p)].$$

Proof. By definition, $\zeta(p)=\sum_{n=1}^{\infty}(1/n^p)$.

$$\begin{aligned}
(\zeta(p))^2 &= \left(\sum_{n=1}^{\infty} \frac{1}{n^p}\right)^2 \\
&= \sum_{m,n=1}^{\infty} \frac{1}{m^p n^p} \\
&= \sum_{n=1}^{\infty} \frac{1}{n^{2p}} + 2\sum_{n,m=1;n>m}^{\infty} \frac{1}{m^p n^p} \\
&= \zeta(2p) + 2\sum_{n,m=1;n>m}^{\infty} \frac{1}{m^p n^p}.
\end{aligned}$$

Rearranging,

$$\begin{aligned}
\frac{1}{2}[(\zeta(p))^2+\zeta(2p)] &= \zeta(2p) + \sum_{n=1}^{\infty}\sum_{m=1}^{n-1} \frac{1}{m^p n^p} \\
&= \sum_{n=1}^{\infty} \frac{1}{n^p n^p} + \sum_{n=1}^{\infty}\sum_{m=1}^{n-1} \frac{1}{m^p n^p} \\
&= \sum_{n=1}^{\infty}\sum_{m=1}^{n} \frac{1}{m^p n^p} \\
&= \sum_{n=1}^{\infty} \frac{1}{n^p}\left(1+\frac{1}{2^p}+\frac{1}{3^p}+\cdots+\frac{1}{n^p}\right),
\end{aligned}$$

as desired.

Gem 82 (CMJ, 24:2, pp. 189–190)

$$\sum_{n=1}^{\infty}\left(1+\frac{1}{2}+\frac{1}{3}+\cdots+\frac{1}{n+1}\right)\frac{1}{n(n+1)}=2.$$

Proof. This is another intriguing result involving the partial sums of the harmonic series. In evaluating this series, we must first recall that

$$\frac{1}{n(n+1)}=\frac{1}{n}-\frac{1}{n+1},$$

so that by telescoping

$$\sum_{n=k}^{\infty} \frac{1}{n(n+1)} = \frac{1}{k}.$$

Taking this into account, and noting that we can freely rearrange the positive terms of the series,

$$\begin{aligned}\sum_{n=1}^{\infty} S &= \left(1 + \frac{1}{2} + \frac{1}{3} + \cdots + \frac{1}{n+1}\right) \frac{1}{n(n+1)} \\ &= \sum_{n=1}^{\infty} \frac{1}{n(n+1)} + \sum_{n=1}^{\infty} \sum_{k=1}^{n} \frac{1}{k+1} \frac{1}{n(n+1)} \\ &= \sum_{n=1}^{\infty} \frac{1}{n(n+1)} + \sum_{k=1}^{n} \frac{1}{k+1} \sum_{n=k}^{\infty} \frac{1}{n(n+1)} \\ &= \sum_{n=1}^{\infty} \frac{1}{n(n+1)} + \sum_{k=1}^{\infty} \frac{1}{k(k+1)} \\ &= 2.\end{aligned}$$

Gem 83 (CMJ, 21:2, p. 151) *For each integer $m > 0$, let $L(m)$ be the number of digits in the decimal representation of m that are at least 5. For example, $L(1) = 0$ and $L(23457126) = 3$. Then*

$$\sum_{t=0}^{\infty} \frac{L(2^t)}{2^t} = \frac{2}{9}.$$

Proof. We decompose L into functions L_k as follows. $L_k(m)$ is 1 if the kth digit of m is at least 5 and 0 otherwise. Then by definition

$$L(m) = \sum_{k=1}^{\infty} L_k(m),$$

where the apparently infinite sum is in fact a finite one since m has only finitely many nonzero digits. Making this substitution,

$$\sum_{t=0}^{\infty} \frac{L(2^t)}{2^t} = \sum_{t=0}^{\infty} \sum_{k=1}^{\infty} \frac{L_k(2^t)}{2^t} = \sum_{k=1}^{\infty} \sum_{t=0}^{\infty} \frac{L_k(2^t)}{2^t}.$$

We must now consider $L_k(m)$. The kth digit of m is precisely the digit in the tenths place of $m/10^k$, so it will be at least 5 exactly if $m/10^k + 0.5$ involves a carry. Indeed we have

$$L_k(m) = \left\lfloor \frac{m}{10^k} + \frac{1}{2} \right\rfloor - \left\lfloor \frac{m}{10^k} \right\rfloor.$$

Using the well-known (and easily proven once-guessed) identity $\lfloor x + 0.5 \rfloor = \lfloor 2x \rfloor - \lfloor x \rfloor$ results in

$$\sum_{t=0}^{\infty} \frac{L_k(2^t)}{2^t} = \sum_{t=0}^{\infty} \left(\frac{1}{2^t} \left\lfloor \frac{2^{t+1}}{10^k} \right\rfloor - \frac{1}{2^{t-1}} \left\lfloor \frac{2^t}{10^k} \right\rfloor \right),$$

which is telescoping and has as its nth partial sum

$$\frac{1}{2^n} \left\lfloor \frac{2^{n+1}}{10^k} \right\rfloor = \frac{1}{2^n} \left(\frac{2^{n+1}}{10^k} - r_n \right),$$

where $0 \leq r_n < 1$. Passing to the limit as $n \to \infty$,

$$\sum_{t=0}^{\infty} \frac{L_k(2^t)}{2^t} = \frac{2}{10^k}.$$

Finally,

$$\sum_{t=0}^{\infty} \frac{L(2^t)}{2^t} = \sum_{t=0}^{\infty} \sum_{k=1}^{\infty} \frac{L_k(2^t)}{2^t} = \sum_{k=1}^{\infty} \frac{2}{10^k} = \frac{2}{9},$$

recognizing the ultimate series as geometric.

Gem 84 (CMJ, 11:1, p. 65) *Let $a > 1$ be an integer. Then*

$$\sum_{n=a}^{\infty} \frac{1}{\binom{n}{a}} = \frac{a}{a-1},$$

where the symbol $\binom{n}{a}$ denotes the binomial coefficient.

Proof. We first make the observation

$$\binom{a+k}{a}^{-1} = \frac{a}{1-a} \left[\binom{a+k}{a-1}^{-1} - \binom{a+k-1}{a-1}^{-1} \right],$$

which follows by straightforward manipulation of the formula

$$\binom{a+k}{a} = \frac{(a+k)!}{a!k!}.$$

Armed with this, we see that the series actually telescopes, and the partial sums have the form

$$\sum_{n=a}^{a+N} \frac{1}{\binom{n}{a}} = \frac{a}{1-a} \left[\binom{a+N}{a-1} - 1 \right],$$

and when we pass to the limit of large N, we get the desired sum.

Gem 85 (CMJ, 30:5, pp. 409–410)

$$\sum_{n=4}^{\infty}\sum_{k=2}^{n-2}\frac{1}{\binom{n}{k}}=\frac{3}{2},$$

where the symbol $\binom{n}{a}$ denotes the binomial coefficient.

Proof. Since all the terms involved are positive, we can interchange the order of summation. (Regard the series as unordered as in Definition 4.1.)

$$\begin{aligned}
\sum_{n=4}^{\infty}\sum_{k=2}^{n-2}\frac{1}{\binom{n}{k}} &= \sum_{k=2}^{\infty}\sum_{n=k+2}^{\infty}\frac{1}{\binom{n}{k}}\\
&= \sum_{k=2}^{\infty}\sum_{n=k+2}^{\infty}\frac{k!(n-k)!}{n!}\\
&= \sum_{k=2}^{\infty}\sum_{n=k+2}^{\infty}\left(\frac{k!(n-k)!}{(k-1)(n-1)!}-\frac{k!(n-k+1)!}{(k-1)n!}\right)\\
&= \sum_{k=2}^{\infty}\left(\frac{2k!}{(k-1)(k+1)!}\right)\\
&= \sum_{k=2}^{\infty}\frac{2}{(k+1)(k-1)}\\
&= \sum_{k=2}^{\infty}\left(\frac{1}{k-1}-\frac{1}{k+1}\right)\\
&= 1+1/2\\
&= 3/2.
\end{aligned}$$

Gem 86 (AMM, 65:6, pp. 452–453) *Let $\{a_i\}_{i=1}^{\infty}$ be integers greater than 1, and define x and y by the infinite series*

$$x=\frac{1}{a_1}+\frac{1}{a_1^2a_2}+\frac{1}{a_1^2a_2^2a_3}+\cdots,$$

$$y=\frac{1}{a_1}+\frac{1}{a_1^3a_2}+\frac{1}{a_1^3a_2^3a_3}+\cdots.$$

Then either x and y are both rational numbers or neither is.

Proof. First, observe that there is no question of convergence of the series defining x and y. Any number of tests will work, perhaps the simplest being the Ratio Test.

Now we prove a simple lemma on inequalities. Let $z = m/n$ be a positive rational number strictly between 0 and 1, and let $e \geq 2$. If the inequality $0 < (za-1)a^{e-1} < 1$ has a positive integral solution a, it cannot have more than one. Suppose instead, if possible, that $a < b$ are two integral solutions, and then

$$\frac{n}{b} < m \leq m(b-a) < \frac{n}{a^{e-1}},$$

which is a contradiction, proving the lemma.

To prove the main claim, let e be an integer not less than 2, and set

$$z_i = a_i^{-1} + (a_i^e a_{i+1})^{-1} + (a_i^e a_{i+1}^e a_{i+2})^{-1} + \cdots.$$

(Note that of course z_i depends on e, but we are suppressing that dependence in the notation.) Then $0 < z_i < 1$ because, term-by-term, the series defining it is less than the series $1/2 + 1/4 + 1/8 + \cdots = 1$. Also $(z_i a_i - 1)a_i^{e-1} = z_{i+1}$. Suppose that z_1 is a rational number and can be written with denominator n. Then it is easy to see by induction using the recursion just given that all the z_i can be written with that same denominator. But there are only finitely many such fractions between 0 and 1, so we must have $z_j = z_{j+k}$ for some $j, k > 0$, and we call this common value z. Then $0 < z_{j+1}, z_{j+k+1} < 1$, so $0 < (za_j - 1)a_j^{e-1} < 1$ and $0 < (za_{j+k} - 1)a_{j+k}^{e-1} < 1$. By the lemma then $a_j = a_{j+k}$. Now, by the natural induction, we have $z_{t+k} = z_t$; $a_{t+k} = a_t$ for all $t \geq j$. Thus the sequence $\{a_n\}$ is eventually periodic. Conversely, if the sequence $\{a_n\}$ is eventually periodic, it is clear that all the z_i are rational. The proof parallels the proof that repeating decimals represent rational numbers in Gem 107. That is, z_1 is rational if and only if the sequence a_n is eventually periodic. This condition, however, does *not* depend on e. Setting $e = 2$ gives $z_1 = x$; setting $e = 3$ gives $z_1 = y$. Hence x and y are both irrational or both rational.

Gem 87 (AMM, 86:1, p. 58) *For each positive integer n, let $f(n)$ be the number of zeros in the decimal representation of n. Choose $a > 0$ and consider the series*

$$\sum_{n=1}^{\infty} \frac{a^{f(n)}}{n^2}.$$

This series converges if $0 < a < 91$ and diverges for $a \geq 91$.

Proof. Consider now only the part of the sum over the $(m+1)$-digit numbers n. That is, $10^m \leq n < 10^{m+1}$. Because our series consists of positive terms, we can group the terms in this way and sum over m without affecting the convergence behavior. For each k, there are $\binom{m}{k}9^{m+1-k}$ numbers with $m+1$ digits that satisfy $f(n) = k$. (Choose which of the m digits other than the first are to be 0, then assign values to the other digits.) For each such n, we have

$$\frac{1}{10^{2m+2}} < \frac{1}{n^2} \leq \frac{1}{10^{2m}}.$$

Using the binomial theorem, this portion of the original series is bounded below by

$$\frac{1}{10^{2m+2}} \sum_{k=0}^{m} \binom{m}{k} 9^{m+1-k} a^k = \frac{9}{100}\left(\frac{9+a}{100}\right)^m$$

and bounded above by

$$\frac{1}{10^{2m}} \sum_{k=0}^{m} \binom{m}{k} 9^{m+1-k} a^k = 9 \left(\frac{9+a}{100} \right)^m .$$

By comparison, then, the original series converges if and only if

$$\sum_{n=0}^{\infty} \left(\frac{9+a}{100} \right)^m$$

does. The geometric series converges for $0 < a < 91$ and diverges for $a \geq 91$.

Gem 88 (AMM, 107:6, p. 568) *For $n > 1$, let $t(n)$ denote the number of unordered factorizations of n into divisors greater than 1, including the trivial factorization with only one factor. For example, $t(12) = 4$ because $12 = 6 \times 2 = 4 \times 3 = 3 \times 2 \times 2$. Then*

$$\sum_{n=2}^{\infty} \frac{t(n)}{n^2} = 1.$$

Proof. This is interesting largely because we obtain this result even though it is virtually impossible to write down a formula for $t(n)$ itself.

Each factorization is completely specified by how many copies of each factor $r \geq 2$ are used. Rearranging this series of positive terms, we can regard it as a sum of copies of $1/n^2$, summing not over integers n but instead over factorizations of integers, so that multiplicity is properly counted.

$$\begin{aligned}
1 + \sum_{n=2}^{\infty} \frac{t(n)}{n^2} &= \prod_{r=2}^{\infty} \left(\sum_{k=0}^{\infty} \frac{1}{r^{2k}} \right) \\
&= \prod_{r=2}^{\infty} \left(\frac{1}{1 - 1/r^2} \right) \\
&= \prod_{r=2}^{\infty} \left(\frac{r^2}{r^2 - 1} \right) \\
&= \prod_{r=2}^{\infty} \left(\frac{r^2}{(r+1)(r-1)} \right) \\
&= \lim_{N \to \infty} \prod_{r=2}^{N} \left(\frac{r^2}{(r+1)(r-1)} \right) \\
&= \lim_{N \to \infty} \frac{2N}{N+1} \\
&= 2,
\end{aligned}$$

and the claim follows.

Gem 89 (AMM, 42:2, pp. 111–112)

$$\frac{1^2}{0!}+\frac{2^2}{1!}+\frac{3^2}{2!}+\frac{4^2}{3!}+\cdots=5e.$$

Proof. Consider the slightly more general series, $\sum_{n=0}^{\infty}(n+k)^2/n!$, where k is any number. The Ratio Test shows that this series converges. To find its sum, we expand as follows:

$$\begin{aligned}
\sum_{n=0}^{\infty}\frac{(n+k)^2}{n!} &= \sum_{n=0}^{\infty}\frac{(n^2+2kn+k^2)}{n!}\\
&= \sum_{n=0}^{\infty}\frac{n^2-n}{n!}+(2k+1)\sum_{n=0}^{\infty}\frac{n}{n!}+k^2\sum_{n=0}^{\infty}\frac{1}{n!}\\
&= \sum_{n=2}^{\infty}\frac{1}{(n-2)!}+(2k+1)\sum_{n=1}^{\infty}\frac{1}{(n-1)!}+k^2\sum_{n=0}^{\infty}\frac{1}{n!}\\
&= \sum_{n=0}^{\infty}\frac{1}{n!}+(2k+1)\sum_{n=0}^{\infty}\frac{1}{n!}+k^2\sum_{n=0}^{\infty}\frac{1}{n!}\\
&= e+(2k+1)e+k^2e.
\end{aligned}$$

Taking $k=1$, we obtain exactly

$$\frac{1^2}{0!}+\frac{2^2}{1!}+\frac{3^2}{2!}+\frac{4^2}{3!}+\cdots=5e.$$

Gem 90 (AMM, 58:2, pp. 116–117)

$$\gamma=\sum_{n=1}^{\infty}\frac{(-1)^n}{n}\left\lfloor\frac{\log n}{\log 2}\right\rfloor,$$

where γ is Euler's constant.

Proof. By definition of γ, we have

$$\begin{aligned}
\gamma &= \lim_{N\to\infty}\left(\sum_{s=1}^{N}\frac{1}{s}-\log N\right)\\
&= \lim_{n\to\infty}\left(\sum_{s=1}^{2^n}\frac{1}{s}-\log 2^n\right).
\end{aligned}$$

Now we focus on the expression in parentheses.

$$\begin{aligned}
\sum_{s=1}^{2^n} \frac{1}{s} - \log 2^n &= \sum_{s=1}^{2^n} \frac{1}{s} - \log 2 - \log 2^{n-1} \\
&= \sum_{s=1}^{2^n} \frac{1}{s} + \sum_{s=1}^{\infty} \frac{(-1)^s}{s} - \log 2^{n-1} \\
&= \sum_{s=2;s \text{ even}}^{2^n} \frac{2}{s} - \log 2^{n-1} + \sum_{s=2^n+1}^{\infty} \frac{(-1)^s}{s} \\
&= \sum_{s=1}^{2^{n-1}} \frac{1}{s} - \log 2^{n-1} + \sum_{s=2^n+1}^{\infty} \frac{(-1)^s}{s}.
\end{aligned}$$

Apply the same procedure to the first two terms of what results to obtain

$$\sum_{s=1}^{2^{n-2}} \frac{1}{s} - \log 2^{n-2} + \sum_{2^{n-1}+1}^{\infty} \frac{(-1)^s}{s} + \sum_{2^n+1}^{\infty} \frac{(-1)^s}{s},$$

and repeat this until the logarithm term is exhausted, leaving us with

$$1 + \sum_{r=1}^{n} \sum_{s=2^r+1}^{\infty} \frac{(-1)^s}{s} = 1 + \sum_{r=1}^{n-1} \left(r \sum_{s=2^r+1}^{2^{r+1}} \frac{(-1)^s}{s} \right) + n \sum_{s=2^n+1}^{\infty} \frac{(-1)^s}{s}.$$

Recall that γ is the limit of this expression as n goes to ∞. However, the last term is the product of n and a tail of the alternating harmonic series with magnitude less than $1/(2^n + 1)$, so the product tends to zero as n grows large. Passing to the limit then,

$$\begin{aligned}
\gamma &= 1 + \sum_{r=1}^{\infty} \left(r \sum_{s=2^r+1}^{2^{r+1}} \frac{(-1)^s}{s} \right) \\
&= 1 + \sum_{s=2}^{\infty} \frac{(-1)^s}{s} \left\{ \frac{\log s}{\log 2} \right\}
\end{aligned}$$

where the curly braces denote the largest integer less than x. This is the same as the greatest integer function except when s is an exact power of 2. To remedy this, write the initial 1 as the sum $1/2 + 1/4 + 1/8 + 1/16 + \cdots$ and combine each of these terms with the corresponding member of the infinite series. This gives exactly the desired

$$\gamma = \sum_{n=1}^{\infty} \frac{(-1)^n}{n} \left\lfloor \frac{\log n}{\log 2} \right\rfloor,$$

once we realize that the $n = 1$ term is zero and can be neglected.

Gem 91 *If $\sum_{n=1}^{\infty} a_n$ is a conditionally convergent series and if r is an arbitrary real number, then there exists a sequence ϵ_n with $\epsilon_n = \pm 1$ such that $\sum_{n=1}^{\infty} \epsilon_n a_n = r$. That is,* *it is possible to redefine the signs of the terms of the series so that it has sum r, leaving the magnitudes of the terms and their order intact.*

Proof. On the surface, this resembles the result in Chapter 3 that a conditionally convergent series has a rearrangement that converges to any given sum. The proof of this result follows the same basic lines. We require only two basic facts about the sequence of magnitudes $b_n = |a_n|$. First, $b_n \to 0$ because the initial series is convergent. Furthermore any tail $\sum_{n=N}^{\infty} b_n$ is divergent because the convergence of the original series is not absolute. (In fact these are the only properties of the sequence $\{a_n\}$ we use. We could have assumed them directly and dispensed with the convergence of $\sum_{n=1}^{\infty} a_n$, but the statement as given seemed in some sense more natural.)

We carry out the construction as follows. Initially, choose $\epsilon_n a_n = b_n$, that is, take the terms positive, until the partial sums first increase above r (if $r < 0$, then we skip this step). Then take ϵ_n so that $\epsilon_n a_n = -b_n$, adding negative terms to decrease the partial sums until they first drop below r. Proceed in this way, alternating between strings of positive and negative terms. Is this always possible? Yes—because every tail $\sum_{n=N}^{\infty} b_n$ diverges, no matter how many terms have already been added, adding (or subtracting) the remaining terms in turn will eventually bring the sum above (or below) r.

Why does this converge to r? Consider first only the partial sums that immediately precede sign changes, which we will call turning points. By construction, such partial sums will differ from r, at most, by the last term added. Since the terms tend to 0, the turning points converge to r. However, any other partial sum must lie between two consecutive turning points, so it is straightforward to check that the sequence of partial sums tends to r, just as claimed.

Gem 92 *Let $\{a_n\}$ be a strictly increasing sequence of positive integers with the property that*

$$\sum_{n=1}^{\infty} \frac{1}{a_n}$$

diverges. Then there are among the prime factors of the various a_n infinitely many distinct primes.

Proof. Suppose for the sake of contradiction that there are only finitely many primes $p_1, p_2, \ldots, p_k$. Then each a_n has the form $\prod {p_i}^{e_i}$. Now consider the series

$$S = \sum_{e_1=0}^{\infty} \sum_{e_2=0}^{\infty} \sum_{e_3=0}^{\infty} \cdots \sum_{e_k=0}^{\infty} \frac{1}{\prod_{i=1}^{k} {p_i}^{e^i}}.$$

On the one hand, this series diverges, because all its terms are positive and it must contain at least all the terms of the divergent series $\sum_{n=1}^{\infty} 1/a_n$. On the other hand, this expression factors as a finite product of convergent geometric series,

$$S = \prod_{i=1}^{k} \left(\sum_{e_i=0}^{\infty} \frac{1}{p_i^{e_i}} \right) = \prod_{i=1}^{k} \left(\frac{p_i}{p_i - 1} \right) < \infty.$$

Gem 93 *Let $a_{m,n} > 0$ be a doubly-indexed sequence of positive terms. Suppose that they are increasing in m in the sense that*

$$a_{m+1,n} \geq a_{m,n}$$

for each m, n. Write $b_n = \lim_{m\to\infty} a_{m,n}$. Also write $A_m = \sum_{n=0}^{\infty} a_{m,n}$, and write $B = \sum_{n=0}^{\infty} b_n$. Then $\lim_{m\to\infty} A_m = B$, with the understanding that any of the limits involved may be infinite. That is,

$$\lim_{m\to\infty} \sum_{n=1}^{\infty} a_{m,n} = \sum_{n=1}^{\infty} \lim_{m\to\infty} a_{m,n}.$$

Proof. This is an example of a circumstance in which we can safely interchange a limit and a summation. In general, situations in which two infinite processes can be interchanged are special, and considerable effort has been dedicated to their study. Here we reduce the situation to more familiar territory by converting the limit to a summation.

Write $c_{m,n} = a_{m,n} - a_{m-1,n}$, where $a_{0,n} = 0$ for each n. Then notice that the $a_{m,n}$ are just the partial sums of the $c_{m,n}$ with respect to m, and use this to rewrite the two expressions that we want to show equal.

$$\lim_{m\to\infty} \sum_{n=1}^{\infty} a_{m,n} = \lim_{m\to\infty} \sum_{n=1}^{\infty} \sum_{k=1}^{m} c_{k,n} = \lim_{m\to\infty} \sum_{k=1}^{m} \sum_{n=1}^{\infty} c_{k,n} = \sum_{k=1}^{\infty} \sum_{n=1}^{\infty} c_{k,n}$$

and also

$$\sum_{n=1}^{\infty} \lim_{m\to\infty} a_{m,n} = \sum_{n=1}^{\infty} \lim_{m\to\infty} \sum_{k=1}^{m} c_{k,n} = \sum_{n=1}^{\infty} \sum_{k=1}^{\infty} c_{k,n}.$$

However, the hypothesis of increasing terms guarantees that each $c_{m,n}$ is nonnegative. Then, we can regard both series in the context of unordered sums, and see that both of these expressions are equal to

$$\sum_{k,n\geq 1} c_{k,n}.$$

Gem 94 *The hypothesis of increasing terms is essential in Gem 93. That is, there exist real numbers $a_{m,n} > 0$ such that, with all notation as before, $\lim_{m\to\infty} A_m \neq B$ even though all limits exist and are finite.*

Proof. Take as a counterexample $a_{n,n} = 1$, and $a_{m,n} = 0$ if $m \neq n$. Then it is easy to see that $A_m = 1$ for every m, but $b_n = 0$ for every n.

We remark that, in light of the way Gem 93 was proven, this gem is just restating the fact that double summations cannot be rearranged freely if the terms are both positive and negative.

Gem 95 (AMM, 63:1, p. 48) *Show that*

$$\frac{1-2^{-2}+4^{-2}-5^{-2}+7^{-2}-8^{-2}+\cdots}{1+2^{-2}-4^{-2}-5^{-2}+7^{-2}+8^{-2}-\cdots}=\frac{2}{3}.$$

Here the numerator consists of terms n^{-2} where n is not a multiple of 3, with alternating signs. The denominator consists of the same terms, but now there is a pair of positive terms, a pair of negative terms, and so on.

Proof. Write N for the numerator and D for the denominator. Then both sums are finite because it is easy to see that convergence is absolute (compare to a p-series).

$$\begin{aligned}N-D&=2(-2^{-2}+4^{-2}-8^{-2}+10^{-2}-14^{-2}+16^{-2}-\cdots)\\&=-2\frac{1}{4}N\\&=\frac{-1}{2}N.\end{aligned}$$

That is, $3N=2D$, so that $N/D=2/3$.

Gem 96 *Assume that the series $\sum_{k=1}^{\infty}a_k$ has both positive and negative terms. Let s_n, p_{n_1}, and q_{n_2} be the nth partial sum, the sum of the positive terms among s_n, and the sum of the negative terms of s_n. ($n=n_1+n_2$). The behavior of $\sum_{k=1}^{\infty}a_n$ can be characterized as follows.*

1. *If $p_{n_1}\to p$ and $q_{n_1}\to q$, the series converges absolutely to $p+q$.*
2. *If $p_{n_1}\to\infty$ and $q_{n_2}\to-\infty$, and also $a_n\to 0$, then the series converges conditionally.*
3. *If $p_{n_1}\to\infty$ and $q_{n_2}\to-\infty$, and also $a_n\not\to 0$, then the series diverges.*
4. *If $p_{n_1}\to\infty$ and $q_{n_2}\to q$, then the series diverges properly to ∞.*
5. *If $p_{n_1}\to p$ and $q_{n_2}\to-\infty$, then the series diverges properly to $-\infty$.*

Proof. If there are only finitely many positive terms or only finitely many negative terms, the matter is trivial. In the first case, we have

$$\sum_{n=1}^{N}|a_n|=p_{n_1}-q_{n_2}<p-q<\infty,$$

so that we have absolute convergence. Since $s_n=p_{n_1}+q_{n_2}$, we know the sum is $p+q$. We also know that $\sum_{n=1}^{\infty}|a_k|=p-q$. The second case was already handled in Chapter 3. In the third case, we apply the converse to the Divergence Theorem. In the fourth case, since eventually $p_{n_1}\to\infty$ and q_{n_2} is bounded below by q, we see that $s_n=p_{n_1}+q_{n_2}$ goes to infinity. The fifth case is analogous.

Gem 97 *A conditionally convergent series cannot be rearranged to form an absolutely convergent series.*

Proof. If $\sum_{n=1}^{\infty} a_n$ could be rearranged to form an absolutely convergent series, then every rearrangement of that series converges to the same sum. This violates the fact that $\sum_{n=0}^{\infty} a_n$, a conditionally convergent series, can be rearranged to sum to any preselected number.

Gem 98 *If $\sum_{n=1}^{\infty} b_n$ diverges and $b_n > 0$, then necessarily*

$$\sum_{n=1}^{\infty} \frac{b_n}{1+b_n}$$

also diverges.

Proof. If b_n does not approach 0, then simple algebra shows that $b_n/(1+b_n)$ does not approach 0 either, so that the second series would have to diverge. If on the other hand b_n approaches 0, then for n large enough, $b_n < 1$, so that $b_n/(1+b_n) > b_n/2$. By the comparison test,

$$\sum_{n=1}^{\infty} \frac{b_n}{1+b_n}$$

diverges.

Gem 99 *If $\sum a_n$ is conditionally convergent, then the terms can be grouped so as to form an absolutely convergent series.*

Proof. Let $\{s_n\}$ be the sequence of partial sums for our series, with $s_n \to s$. Choose an increasing sequence $0 = n_0 < n_1 < n_2 < n_3 < \cdots$ such that for $k > 0$,

$$s - \frac{1}{2^k} < s_{n_k} < s + \frac{1}{2^k}$$

and group the series as $b_1 + b_2 + b_3 + \cdots$, where the b_i are

$$b_i = \sum_{n=n_{i-1}+1}^{n_i} a_n.$$

Notice that $b_i = s_{n_i} - s_{n_{i-1}}$, so that for $i > 1$,

$$|b_i| \leq |s_{n_i} - s| + |s_{n_{i-1}} - s| < \frac{1}{2^i} + \frac{1}{2^{i-1}} = \frac{3}{2^i}.$$

Then the terms $|b_i|$ are dominated by the terms of a convergent geometric series, and so $\sum b_i$ is an absolutely convergent grouping of our series.

Gem 100 *If $\sum_{n=1}^{\infty} c_n$ is a convergent series of positive terms, then*

$$\sum_{n=1}^{\infty} \frac{\sqrt{c_n}}{n}$$

converges.

Proof. By Cauchy's Inequality,

$$\left(\sum_{n=1}^{\infty}\frac{\sqrt{c_n}}{n}\right)^2 \leq \left(\sum_{n=1}^{\infty} c_n\right)\left(\sum_{n=1}^{\infty}\frac{1}{n^2}\right).$$

Since both series on the right converge, so also does the series on the left.

Gem 101 *If $\sum_{n=1}^{\infty} c_n$ is a convergent series of positive terms, then*

$$\sum_{n=1}^{\infty}\sqrt{\frac{c_n}{n}}$$

may converge or diverge.

Proof. A convergent example would come from $c_n = 1/n^3$, which would give

$$\sum_{n=1}^{\infty}\sqrt{\frac{c_n}{n}} = \sum_{n=1}^{\infty}\frac{1}{n^2}.$$

A divergent example would come from $c_n = 1/n(\log n)^2$, which would give

$$\sum_{n=2}^{\infty}\sqrt{\frac{c_n}{n}} = \sum_{n=2}^{\infty}\frac{1}{n\log n},$$

where we appeal to the Abel-Dini scale to diagnose these series.

Gem 102 *For each real number x,*

$$\sum_{n=1}^{\infty}\frac{\sin(nx)}{n}$$

converges.

Proof. If x is an integer multiple of π, the series is identically 0 and there is nothing to show. Assume that $x \neq 0$ for the sequel.

Let $a_n = \sin(nx)$ and $b_n = 1/n$ and appeal to the result by Dirichlet as given in Corollary 2.23.1.

Key to our argument is the trigonometric identity

$$2\sin(x/2)\sin(nx) = \cos(n-1/2)x - \cos(n+1/2)x.$$

Let $s_n = \sum_{k=1}^{n} a_k$, assume x is not an integer multiple of π, and use telescoping to see that

$$s_n = \frac{1}{2}\frac{\cos(x/2) - \cos[(n+1/2)x]}{\sin(x/2)} = \frac{1}{2}\left[\frac{\cos(x/2)}{\sin(x/2)} - \frac{\cos[(n+1/2)x]}{\sin(x/2)}\right].$$

Then the partial sums s_n are bounded. Since $\{b_n\}$ is a monotone sequence tending to 0, the Dirichlet result guarantees

$$\sum_{n=1}^{\infty} a_n b_n = \sum_{n=1}^{\infty} \frac{\sin(nx)}{n}$$

converges.

Gem 103 *If $\sum_{n=1}^{\infty} c_n$ converges, then $\sum_{n=1}^{\infty} c_n \sin(n)$ may converge or diverge.*

Proof. If $\sum_{n=1}^{\infty} c_n$ is absolutely convergent, it is immediate to see that $|c_n \sin n| < |c_n|$, so that $\sum_{n=1}^{\infty} c_n \sin(n)$ is still absolutely convergent.

The argument for divergence is delicate. Now we begin with a conditionally convergent series $\sum_{k=1}^{\infty}(-1)^{k+1}a_k$ with $a_k > 0$. Identify a collection of subintervals of the real line,

$$I_k = \left[(k-1)\pi + \frac{\pi}{4}, (k-1)\pi + \frac{3\pi}{4}\right].$$

Each interval I_k has length $\pi/2 > 1$, so it contains at least one integer point n_k. Note therefore that $|\sin(n_k)| \geq \frac{\sqrt{2}}{2}$. Define $c_n = 0$ if n is not one of the n_k and $c_{n_k} = (-1)^{k+1}a_k$ if $n = n_k$ in order to create a series $\sum_{n=1}^{\infty} c_n$ that converges. We also have

$$(-1)^{k+1}\sin(n_k) \geq \frac{\sqrt{2}}{2}$$

from which follows

$$c_{n_k}\sin(n_k) = a_k \sin(n_k) \geq \frac{\sqrt{2}}{2}a_k > 0.$$

Since $\sum_{k=1}^{\infty} a_k$ diverges it follows by the Comparison Test, that $\sum c_n \sin(n)$ diverges.

Gem 104

$$\sqrt{2}\sqrt[4]{4}\sqrt[8]{8}\cdots\sqrt[2^n]{2^n}\cdots = 4.$$

Proof. We take logarithms, reducing this infinite product to an infinite series,

$$\log\left(\prod_{n=1}^{\infty}\sqrt[2^n]{2^n}\right) = \sum_{n=1}^{\infty}\left(\log\sqrt[2^n]{2^n}\right) = \sum_{n=1}^{\infty}\left(\frac{n}{2^n}\log 2\right) = \log 2\left(\sum_{n=1}^{\infty}\frac{n}{2^n}\right).$$

By Gem 42, this series sums to 2, and we have

$$\log\left[\prod_{n=1}^{\infty}\left(\sqrt[2^n]{2^n}\right)\right] = 2\log 2.$$

Exponentiating gives exactly what is desired.

Gem 105

$$\sum_{n=1}^{\infty} (\sqrt[n]{n} - 1) = \infty.$$

Proof. We appeal to Gem 16. It is sufficient to show that

$$\prod_{n=1}^{\infty} \sqrt[n]{n} = \infty.$$

Taking logarithms, we see

$$\log \prod_{n=1}^{\infty} \sqrt[n]{n} = \sum_{n=1}^{\infty} \log \sqrt[n]{n} = \sum_{n=1}^{\infty} \frac{\log n}{n} \geq \sum_{n=1}^{\infty} \frac{1}{n}$$

and we are done by comparison with the harmonic series.

Gem 106

$$\sum_{n=1}^{\infty} \frac{\cot^{-1}(1+n+n^2)}{\tan^{-1}(n)}$$

converges.

Proof. First, recalling the appropriate trigonometric identities,

$$\cot^{-1}(1+n+n^2) = \tan^{-1} \frac{1}{1+n+n^2} = \tan^{-1} \frac{(n+1)-n}{1+(n+1)n} = \tan^{-1}(n+1) - \tan^{-1}(n).$$

By this substitution,

$$\sum_{n=1}^{\infty} \frac{\cot^{-1}(1+n+n^2)}{\tan^{-1}(n)} = \sum_{n=1}^{\infty} \frac{\tan^{-1}(n+1) - \tan^{-1}(n)}{\tan^{-1}(n)} = \sum_{n=1}^{\infty} \left(\frac{\tan^{-1}(n+1)}{\tan^{-1}(n)} - 1 \right).$$

Appealing to Gem 16, it will suffice to show that

$$\prod_{n=1}^{\infty} \frac{\tan^{-1}(n+1)}{\tan^{-1}(n)}$$

is convergent, but of course this product telescopes. We have

$$\prod_{n=1}^{N} \frac{\tan^{-1}(n+1)}{\tan^{-1}(n)} = \frac{\tan^{-1}(N+1)}{\tan^{-1} 1}.$$

Since $\lim_{n\to\infty} \tan^{-1} n = \pi/2$,

$$\prod_{n=1}^{\infty} \frac{\tan^{-1}(n+1)}{\tan^{-1}(n)} = \frac{\pi/2}{\pi/4} = 2 < \infty.$$

Gem 107 *Let x be a repeating decimal between* 0 *and* 1, *so that it has the decimal representation*

$$x = 0.a_1a_2a_3\dots a_kb_1b_2b_3\dots b_lb_1b_2b_3\dots b_lb_1b_2b_3\dots b_l\dots$$

for some digits a_i, b_j. Then x is rational and can be written

$$x = \frac{a_1a_2a_3\dots a_kb_1b_2b_3\dots b_l - a_1a_2a_3\dots a_k}{99\dots900\dots0}$$

where there are l 9's and k 0's, and all the expressions are interpreted as forming numbers by concatenating digits, not as products. For example

$$0.12456456456\dots = \frac{12456 - 12}{99900}.$$

Proof. Let x have the form described. Then we have

$$10^{k+l}x = a_1a_2a_3\dots a_kb_1b_2b_3\dots b_l.b_1b_2b_3\dots b_lb_1b_2b_3\dots b_lb_1b_2b_3\dots b_l\dots,$$
$$10^kx = a_1a_2a_3\dots a_k.b_1b_2b_3\dots b_lb_1b_2b_3\dots b_lb_1b_2b_3\dots b_l\dots.$$

Notice that both decimals have the same repeating part, so we can subtract to find

$$(10^{k+l} - 10^k)x = a_1a_2a_3\dots a_kb_1b_2b_3\dots b_l - a_1a_2a_3\dots a_k.$$

We now simply solve for x, noticing that $10^{k+l} - 10^k = 99\dots900\dots0$, where there are l 9's and k 0's.

5

Series and the Putnam Competition

The William Lowell Putnam Mathematical Competition is the premier competition among undergraduate students in mathematics. This contest, held over one entire day each year, is designed to test students' powers of creative mathematical thinking, problem solving, and knowledge of basic mathematics. There are at least two purposes for this chapter. First, the large number of problems about real infinite series, on the surface a small part of mathematics, included historically on the Putnam can be seen as a rough indication of their importance. Second and more importantly, this is an opportunity for the reader to see how the theory of infinite series can be applied to specific problems. The reader has by this point seen a great deal of theorems and background material about infinite series without necessarily seeing any particular application of the theorems. We wish for the reader to have experience solving unfamiliar problems in the context of infinite series. While it is true that many of the gems in Chapter 4 have this form, in that section of the book theory and application appear all mixed together, with the distinctions subjective and implicit. In this chapter we wish to be more explicit. Here we provide the reader with a collection of problems written not by the authors of this text but by a collection of eminent mathematicians, and we extend an invitation for the reader to challenge himself or herself to solve them. For this reason, we list the problems twice—first without their solutions, and then again with solutions and other commentary.

In compiling this chapter, we have relied extensively on the three volumes of *The William Lowell Putnam Mathematical Competition* in the MAA Problem Books series. The first volume was edited by A.M. Gleason, R.E. Greenwood, L.M. Kelly; the second by G.L. Alexanderson, L.F. Klosinski, and L.C. Larson; and the third by K.S. Kedlaya, B. Poonen, and R. Vakil. These books are excellent, and we are indebted to what they provide. The reader who is interested in mathematical problem solving in a more general context would do well to study these books. We have compiled those problems that relate to the material in this book. Actually this is not precisely true; though this book has briefly treated power series, we do not consider this topic core to the book. The Putnam Competition has featured a wealth of problems on power series. To the interested reader, we once more heartily recommend the MAA Problem Books. In all cases the problem statement has been taken directly from those versions, and in almost all cases the solutions have as well. What material we have taken from these books appears in virtually the same form as the originals; when changes have been made, they are typographical in nature. It is always explicitly stated when the material given as a solution is not due directly to our sources. Conversely, when commentary follows a solution, it is due to the authors of this book and these exceptions are clearly marked. Each problem and solution is labelled with an ordered triple $[x, y, z]$ where the first number represents the number of the Putnam Competition in which the problem was given, the second entry gives the year of that contest, and the third entry gives the problem number within the contest.

The reader will notice that we generally do not give solutions in the same style of the proofs of our theorems in earlier chapters of this book. The solutions are somewhat in a conversational style as may be typical of the everyday life of practicing mathematicians or students who participate in mathematical competitions. The standard results are referred to by familiar name, description, or sometimes implicit use, rather than by "official" number as given in this presentation. This should be in no way intimidating, however; in some cases when we use a particularly less well-known result, we refer the reader back to the appropriate place in the book, if a refresher is desired. We hope this choice will help the reader make the transition from learning the subject to applying the subject as a mathematician (armchair or otherwise). By this point, the reader has already read at least one entire book (almost) on infinite series. In this sendoff, we are discussing infinite series as mature mathematicians, and we treat the reader as knowledgeable.

5.1 The Problems

Problem [3, 1940, 7]

If $u_1^2 + u_2^2 + u_3^2 + \cdots$ and $v_1^2 + v_2^2 + v_3^2 + \cdots$ are convergent series of real constants, prove that

$$(u_1 - v_1)^p + (u_2 - v_2)^p + (u_3 - v_3)^p + \cdots,$$

p an integer ≥ 2, is convergent.

Problem [5, 1942, 3]

Is the following series convergent or divergent?

$$1 + \frac{1}{2} \cdot \frac{19}{7} + \frac{2!}{3^2} \cdot \left(\frac{19}{7}\right)^2 + \frac{3!}{4^3} \cdot \left(\frac{19}{7}\right)^3 + \frac{4!}{5^4} \cdot \left(\frac{19}{7}\right)^4 + \cdots$$

Problem [8, 1948, A3]

Let $\{a_n\}$ be a decreasing sequence of positive numbers with limit 0 such that

$$b_n = a_n - 2a_{n+1} + a_{n+2} \geq 0$$

for all n. Prove that

$$\sum_{n=1}^{\infty} nb_n = a_1.$$

Problem [9, 1949, A3]

Assume that the complex numbers $a_1, a_2, \ldots, a_n, \ldots$ are all different from zero, and that $|a_r - a_s| > 1$ for $r \neq s$. Show that the series

$$\sum_{n=1}^{\infty} \frac{1}{a_n^3}$$

converges.

Problem [9, 1949, B2(i)]

Prove that

$$\sum_{n=2}^{\infty} \frac{\cos(\log \log n)}{\log n}$$

diverges.

Problem [10, 1950, A2]

Test for convergence the series

$$\frac{1}{\log(2!)} + \frac{1}{\log(3!)} + \frac{1}{\log(4!)} + \cdots + \frac{1}{\log(n!)} + \cdots,$$

$$\frac{1}{3} + \frac{1}{3\sqrt{3}} + \frac{1}{3\sqrt{3}\sqrt[3]{3}} + \cdots + \frac{1}{3\sqrt{3}\sqrt[3]{3}\cdots\sqrt[n]{3}} + \cdots.$$

Problem [11, 1951, A7]

Show that if the series $a_1 + a_2 + a_3 + \cdots + a_n + \cdots$ converges, then the series $a_1 + a_2/2 + a_3/3 + \cdots a_n/n + \cdots$ converges also.

Problem [12, 1952, B5]

If the terms of a sequence a_n are monotonic, and if $\sum_{n=1}^{\infty} a_n$ converges, show that $\sum_{n=1}^{\infty} n(a_n - a_{n+1})$ converges.

Problem [13, 1953, B1]

Is the infinite series

$$\sum_{n=1}^{\infty} \frac{1}{n^{(n+1)/n}}$$

convergent? Prove your statement.

Problem [14, 1954, A6]

Suppose that $u_0, u_1, u_2, \ldots$ is a sequence of real numbers such that

$$u_n = \sum_{k=1}^{\infty} u_{n+k}^2 \quad \text{for } n = 0, 1, 2, \ldots.$$

Prove that if $\sum u_n$ converges, then $u_k = 0$ for all k.

Problem [14, 1954, B2]

Assume as known the (true) fact that the alternating harmonic series

$$1 - 1/2 + 1/3 - 1/4 + 1/5 - 1/6 + 1/7 - 1/8 + \cdots$$

is convergent, and denote its sum by s. Rearrange the series as follows,

$$1 + 1/3 - 1/2 + 1/5 + 1/7 - 1/4 + 1/9 + 1/11 - 1/6 + \cdots.$$

Assume as known the (true) fact that this series is also convergent, and denote its sum by S. Denote by s_k, S_k the kth partial sum of the respective series. Prove the following statements.

1. $S_{3n} = s_{4n} + \frac{1}{2}s_{2n}$.
2. $S \neq s$.

Problem [15, 1955, A3]

Suppose that $\sum_{i=1}^{\infty} x_i$ is a convergent series of positive terms that monotonically decrease (that is, $x_1 \geq x_2 \geq x_3 \geq \cdots$). Let P denote the set of all numbers that are sums of some (finite or infinite) subseries of $\sum_{n=1}^{\infty} x_i$. Show that P is an interval if and only if

$$x_n \leq \sum_{i=n+1}^{\infty} x_i \quad \text{for every integer } n,$$

which we call condition (1).

Problem [16, 1956, B6]

Given $T_1 = 2$, $T_{n+1} = T_n^2 - T_n + 1$, $n > 0$, prove:

1. If $m \neq n$, T_m and T_n have no common factor greater than 1.
2. $\sum_{i=1}^{\infty} \frac{1}{T_i} = 1$.

Problem [18, 1958, A6]

What is the smallest amount that may be invested at interest rate i, compounded annually, in order that one may withdraw 1 dollar at the end of the first year, 4 dollars at the end of the second year, ... , n^2 dollars at the end of the nth year, in perpetuity.

Problem [19, 1958, B5]

The lengths of successive segments of a broken line are represented by the successive terms of the harmonic progression $1, 1/2, 1/3, \dots, 1/n, \dots$. Each segment makes with the preceding segment a given angle θ. What is the distance and what is the direction of the limiting point (if there is one) from the initial point of the first segment?

Problem [20, 1959, B2]

Let c be a positive real number. Prove that c can be expressed in infinitely many ways as a sum of infinitely many distinct terms selected from the sequence

$$1/10, 1/20, \dots, 1/10n, \dots.$$

Problem [21, 1960, B2]

Evaluate the double series

$$\sum_{j=0}^{\infty}\sum_{k=0}^{\infty} 2^{-3k-j-(k+j)^2}.$$

Problem [21, 1960, B6]

Any positive integer may be written in the form $n = 2^k(2l + 1)$. Let $a_n = e^{-k}$ and $b_n = a_1 a_2 a_3 \cdots a_n$. Prove that $\sum b_n$ converges.

Problem [24, 1963, B5]

Let $\{a_n\}$ be a sequence of real numbers satisfying the inequalities

$$0 \leq a_k \leq 100a_n \quad \text{for } n \leq k \leq 2n \quad \text{and} \quad n = 1, 2, \dots,$$

and such that the series

$$\sum_{n=0}^{\infty} a_n$$

converges. Prove that

$$\lim_{n \to \infty} n a_n = 0.$$

Problem [25, 1964, A3]

Let $P_1, P_2, \ldots$ be a sequence of distinct points that is dense in the interval $(0, 1)$. The points $P_1, P_2, \ldots, P_{n-1}$ decompose the interval into n parts, and P_n decomposes one of these into two parts. Let a_n and b_n be the lengths of these two intervals. Prove that

$$\sum_{n=1}^{\infty} a_n b_n (a_n + b_n) = 1/3.$$

(A sequence of points in an interval is said to be dense when every subinterval contains at least one point of the sequence.)

Problem [25, 1964, A5]

Prove that there is a constant K such that the following inequality holds for any sequence of positive numbers $a_1, a_2, a_3, \ldots$:

$$\sum_{n=1}^{\infty} \frac{n}{a_1 + a_2 + \cdots + a_n} \leq K \sum_{n=1}^{\infty} \frac{1}{a_n}.$$

Problem [25, 1964, B1]

Let u_k $(k = 1, 2, \ldots)$ be a sequence of integers, and let V_n be the number of those that are less than or equal to n. Show that if

$$\sum_{k=1}^{\infty} 1/u_k < \infty,$$

then

$$\lim_{n \to \infty} V_n / n = 0.$$

Problem [25, 1964, B5]

Let u_n $(n = 1, 2, 3, \ldots)$ denote the least common multiple of the first n terms of a strictly increasing sequence of positive integers (for example, the sequence $1, 2, 3, 4, 5, 6, 10, 12$). Prove that the series

$$\sum_{n=1}^{\infty} 1/u_n$$

is convergent.

Problem [27, 1966, B3]

Show that if the series

$$\sum_{n=1}^{\infty} \frac{1}{p_n}$$

is convergent, where $p_1, p_2, p_3, \ldots, p_n, \ldots$ are positive real numbers, then the series

$$\sum_{n=1}^{\infty} \frac{n^2}{(p_1 + p_2 + \cdots + p_n)^2} p_n$$

is also convergent.

Problem [30, 1969, B5]

Let $a_1 < a_2 < a_3 < \cdots$ be an increasing sequence of positive integers. Let the series $\sum_{n=1}^{\infty} 1/a_n$ be convergent. For any number x let $k(x)$ be the number of the a_n's that do not exceed x. Show that $\lim_{x\to\infty} k(x)/x = 0$.

Problem [36, 1975, B5]

Let $f_0(x) = e^x$ and $f_{n+1}(x) = x f_n'(x)$ for $n = 0, 1, 2, \ldots$. Show that

$$\sum_{n=0}^{\infty} \frac{f_n(1)}{n!} = e^e.$$

Problem [39, 1978, B2]

Express

$$\sum_{m=1}^{\infty} \sum_{n=1}^{\infty} \frac{1}{m^2 n + m n^2 + 2mn}$$

as a rational number.

Problem [43, 1982, A2]

For positive real x, let

$$B_n(x) = 1^x + 2^x + 3^x + \cdots + n^x.$$

Prove or disprove the convergence of

$$\sum_{n=2}^{\infty} \frac{B_n(\log_n 2)}{(n \log_2 n)^2}.$$

Problem [45, 1984, A2]

Express $\sum_{k=1}^{\infty}(6^k/(3^{k+1}-2^{k+1})(3^k-2^k))$ as a rational number.

Problem [47, 1986, A3]

Evaluate $\sum_{n=0}^{\infty}$ Arccot (n^2+n+1), where Arccot t for $t \geq 0$ denotes the number θ in the interval $0 < \theta \leq \pi/2$ with $\cot\theta = t$.

Problem [48, 1987, A6]

For each positive integer n, let $a(n)$ be the number of zeroes in the base 3 representation of n. For which positive real numbers x does the series

$$\sum_{n=1}^{\infty}\frac{x^{a(n)}}{n^3}$$

converge?

Problem [49, 1988, A3]

Determine, with proof, the set of real numbers x for which

$$\sum_{n=0}^{\infty}\left(\frac{1}{n}\csc\frac{1}{n}-1\right)^x$$

converges.

Problem [49, 1988, B4]

Prove that if $\sum_{n=1}^{\infty} a_n$ is a convergent series of positive real numbers, then so is

$$\sum_{n=1}^{\infty}(a_n)^{n/(n+1)}.$$

Problem [55, 1994, A1]

Suppose that a sequence $a_1, a_2, a_3, \ldots$ satisfies $0 < a_n < a_{2n} + a_{2n+1}$ for all $n \geq 1$. Prove that series $\sum_{n=1}^{\infty} a_n$ diverges.

Problem [60, 1999, A4]

Sum the series

$$\sum_{m=1}^{\infty}\sum_{n=1}^{\infty}\frac{m^2 n}{3^m(n3^m+m3^n)}.$$

Problem [61, 2000, A1]

Let A be a positive real number. What are the possible values of $\sum_{j=0}^{\infty} x_j^2$, given that $x_0, x_1, \ldots$ are positive numbers for which $\sum_{j=0}^{\infty} x_j = A$?

Problem [62, 2001, B3]

For any positive integer n, let $\langle n \rangle$ denote the closest integer to $\sqrt{n}$. Evaluate

$$\sum_{n=1}^{\infty} \frac{2^{\langle n \rangle} + 2^{-\langle n \rangle}}{2^n}.$$

Problem [63, 2002, A6]

Fix an integer $b \geq 2$. Let $f(1) = 1$, $f(2) = 2$, and for each $n \geq 3$, define $f(n) = nf(d)$, where d is the number of base-b digits of n. For which values of b does

$$\sum_{n=1}^{\infty} \frac{1}{f(n)}$$

converge?

5.2 The Solutions

Problem [3, 1940, 7]

If $u_1^2 + u_2^2 + u_3^2 + \cdots$ and $v_1^2 + v_2^2 + v_3^2 + \cdots$ are convergent series of real constants, prove that

$$(u_1 - v_1)^p + (u_2 - v_2)^p + (u_3 - v_3)^p + \cdots,$$

p an integer ≥ 2, is convergent.

Solution. Let $A = u_1^2 + u_2^2 + u_3^2 + \cdots$ and $B = v_1^2 + v_2^2 + v_3^2 + \cdots$. Since

$$(u_i + v_i)^2 + (u_i - v_i)^2 = 2u_i^2 + 2v_i^2$$

we have, for any positive integer n,

$$\sum_{i=1}^{n}(u_i - v_i)^2 \leq 2\sum_{i=1}^{n} u_i^2 + 2\sum_{i=1}^{n} v_i^2 \leq 2A + 2B.$$

Since the terms are all nonnegative., it follows that

$$\sum_{i=1}^{\infty}(u_i - v_i)^2 \text{ is convergent.}$$

Therefore, the terms approach zero, so there exists an integer k so that

$$(u_i - v_i)^2 < 1 \text{ for all } i \geq k.$$

If p is an integer and $p \geq 2$, then $|u_i - v_i|^p \leq (u_i - v_i)^2$ for all $i \geq k$, so the series

$$\sum_{i=1}^{\infty}(u_i - v_i)^p$$

is absolutely convergent, and therefore convergent.

Remarks. The only reason that p needs to be an integer is so that the expression will be defined even when $u_i - v_i < 0$. If we use the series $\sum_{i=1}^{\infty} |u_i - v_i|^p$, the result will hold even if p is not an integer.

A related generalization of this problem would include complex numbers. While strictly speaking we only defined infinite series of real numbers in Chapter 1, we have already noted that all the same definitions make sense in the complex numbers. In this space it makes sense to raise any real number, positive or negative, to any complex power. The same proof given above actually establishes the following:

If $u_1^2 + u_2^2 + u_3^2 + \cdots$ and $v_1^2 + v_2^2 + v_3^2 + \cdots$ are convergent series of real constants (that is, the u_i and v_i are real) and p is any complex number with real part not less than 2, prove that

$$(u_1 - v_1)^p + (u_2 - v_2)^p + (u_3 - v_3)^p + \cdots,$$

is absolutely convergent.

Problem [5, 1942, 3]

Is the following series convergent or divergent?

$$1 + \frac{1}{2} \cdot \frac{19}{7} + \frac{2!}{3^2} \cdot \left(\frac{19}{7}\right)^2 + \frac{3!}{4^3} \cdot \left(\frac{19}{7}\right)^3 + \frac{4!}{5^4} \cdot \left(\frac{19}{7}\right)^4 + \cdots.$$

Solution. Use the Ratio Test. Let

$$a_n = \frac{(n-1)!}{n^{n-1}} \left(\frac{19}{7}\right)^{n-1}, a_{n+1} = \frac{n!}{(n+1)^n} \left(\frac{19}{7}\right)^n.$$

Then

$$R_n = \frac{a_{n+1}}{a_n} = \frac{n^n}{(n+1)^n} \frac{19}{7} = \frac{1}{\left(1 + \frac{1}{n}\right)^n} \frac{19}{7},$$

and

$$\lim_{n \to \infty} R_n = \frac{19}{7} \lim_{n \to \infty} \frac{1}{\left(1 + \frac{1}{n}\right)^n} = \frac{19}{7} \frac{1}{e}.$$

Since $19/7 < 2.715$ and $e > 2.718$, $19/(7e) < 1$ and the series converges.

Problem [8, 1948, A3]

Let $\{a_n\}$ be a decreasing sequence of positive numbers with limit 0 such that

$$b_n = a_n - 2a_{n+1} + a_{n+2} \geq 0$$

for all n. Prove that

$$\sum_{n=1}^{\infty} nb_n = a_1.$$

Solution. Since the b_n's are the second differences of the a_n's, it is convenient to let $c_n = a_n - a_{n+1}$; then

$$c_n - c_{n+1} = a_n - 2a_{n+1} + a_{n+2} = b_n.$$

Since the a_n's decrease to zero, $c_n \geq 0$ for all n, and $c_n \to 0$.

For $k \geq m$, we have $\sum_{i=m}^{k} b_i = c_m - c_{k+1}$, and therefore

$$\sum_{i=m}^{\infty} b_i = c_m = a_m - a_{m+1}.$$

Similarly,

$$\sum_{m=1}^{k}(a_m - a_{m+1}) = a_1 - a_{k+1}, \quad \text{so} \quad \sum_{m=1}^{\infty}(a_m - a_{m+1}) = a_1.$$

Thus

$$\sum_{m=1}^{\infty}\left(\sum_{i=m}^{\infty} b_i\right) = a_1.$$

The b_n's are nonnegative, and when summing nonnegative terms, rearrangement does not affect the value of the sum. For each index n, the term b_n appears exactly n times in the preceding double sum, once in each of the terms $\sum_{i=m}^{\infty} b_i$ for $m = 1, 2, \ldots, n$. Hence

$$\sum_{n=1}^{\infty} nb_n = \sum_{m=1}^{\infty}\left(\sum_{i=m}^{\infty} b_i\right) = a_1.$$

Remark. We noted in Chapter 2 that taking first differences is dual to taking partial sums. Since the b's are the second differences of the a's, the partial sums of the b's will be (almost) the first differences of the a's. Specifically,

$$\sum_{i=1}^{k} b_i = a_1 - a_2 - a_{k+1} + a_{k+2}.$$

Furthermore, the other factor, n, can be viewed as the partial sums of $\sum 1$, so it will have very nice first differences $n - (n+1) = -1$. This suggests that Abel's "summation by parts" formula (see Chapter 2) could improve matters, which is true. We do not do the calculation here, because it is essentially the same as the given solution, though perhaps more motivated. Indeed, the interested reader can try it for a bit of practice in working through the details when using Abel. In either case, the key to seeing the solution is recognizing that the b's are the second differences of the a's.

Problem [9, 1949, A3]

Assume that the complex numbers $a_1, a_2, \ldots, a_n, \ldots$ are all different from zero, and that $|a_r - a_s| > 1$ for $r \neq s$. Show that the series

$$\sum_{n=1}^{\infty} \frac{1}{a_n^3}$$

converges.

Solution. Let $S_k = \{n : k < |a_n| \leq k+1\}$ for $k = 0, 1, 2, \ldots$. The discs $|z - a_n| \leq \frac{1}{2}$ are all disjoint by hypothesis, and for $n \in S_k$ these discs all lie in the annulus

$$\left\{z : k - \frac{1}{2} \leq |z| \leq k + \frac{3}{2}\right\}$$

(a disc if $k = 0$). Let the cardinality of the set S_k be denoted by $|S_k|$. Then adding areas gives

$$|S_k|\frac{\pi}{4} \leq \pi\left[\left(k + \frac{3}{2}\right)^2 - \left(k - \frac{1}{2}\right)^2\right] = 2\pi(2k+1)$$

so that $|S_k| \leq 8(2k+1)$ for $k > 0$. A separate calculation shows that $|S_0| \leq 9$.

Then

$$\sum_{n \in S_k} \frac{1}{|a_n|^3} \leq \frac{|S_k|}{k^3} \leq \frac{8(2k+1)}{k^3} \leq \frac{24}{k^2}$$

for $k \geq 1$ because $2k + 1 \leq 3k$. Since S_0 is finite,

$$\sum_{n \in S_0} \frac{1}{|a_n|^3}$$

is finite.

Hence we have

$$\sum_{n=1}^{\infty} \frac{1}{|a_n|^3} = \sum_{k=0}^{\infty} \sum_{n \in S_k} \frac{1}{|a_n|^3} \leq \sum_{n \in S_0} \frac{1}{|a_n|^3} + \sum_{k=1}^{\infty} \frac{24}{k^2} < \infty.$$

The rearrangement of the sum in the first step is permissible since the terms are all positive. Thus the original series converges absolutely.

Problem [9, 1949, B2(i)]

Prove that

$$\sum_{n=2}^{\infty} \frac{\cos(\log\log n)}{\log n}$$

diverges.

Solution. A convergent series cannot have blocks of terms whose sum is arbitrarily large. We shall show that the given series has such blocks.

For a positive integer k consider the set N_k of integers n such that

$$2\pi k - \frac{1}{3}\pi \leq \log\log n \leq 2\pi k$$

and let

$$T_k = \sum_{N_k} \frac{\cos(\log\log n)}{\log n}.$$

We shall prove that $T_k \to \infty$ as $k \to \infty$. Now $N_k = \{n : \exp(\exp(2\pi k - \frac{1}{3}\pi)) \leq n \leq \exp(\exp(2\pi k))\}$ and therefore $|N_k|$, the number of elements in N_k, satisfies

$$|N_k| \geq \exp(\exp(2\pi k)) - \exp(\alpha \exp(2\pi k)) - 1$$

where $\alpha = \exp(-\frac{1}{3}\pi)$.

Each term in the sum for T_k is at least

$$\frac{\cos\left(-\frac{1}{3}\pi\right)}{\exp(2\pi k)} = \frac{1}{2\exp(2\pi k)}$$

so

$$T_k \geq \frac{1}{2x_k}(\exp(x_k) - \exp(\alpha x_k) - 1)$$

where $x_k = \exp(2\pi k)$.

Now

$$\lim_{x\to\infty} \frac{1}{x}(\exp x - \exp \alpha x) = \lim_{x\to\infty} (\exp x - \alpha \exp \alpha x) = \infty$$

by L'Hôpital's Rule, using the fact that $\alpha < 1$. Since $x_k \to \infty$ as $k \to \infty$, it follows that $T_k \to \infty$ as $k \to \infty$, and this proves that the given series diverges.

Problem [10, 1950, A2]

Test for convergence the series

$$\frac{1}{\log(2!)} + \frac{1}{\log(3!)} + \frac{1}{\log(4!)} + \cdots + \frac{1}{\log(n!)} + \cdots$$

and

$$\frac{1}{3}+\frac{1}{3\sqrt{3}}+\frac{1}{3\sqrt{3}\sqrt[3]{3}}+\cdots+\frac{1}{3\sqrt{3}\sqrt[3]{3}\cdots\sqrt[n]{3}}+\cdots.$$

Solution. For $n \geq 2$, we have $n^n > n!$, hence $n\log n > \log(n!)$ and

$$\frac{1}{\log(n!)} > \frac{1}{n\log n}.$$

The first series therefore dominates the series

$$\sum_{n=2}^{\infty}\frac{1}{n\log n}.$$

Since

$$\int_2^x \frac{dt}{t\log t} = \log\log x - \log\log 2,$$

the improper integral $\int_2^\infty dt/(t\log t)$ diverges, and hence, by the integral test, so does $\sum_{n=2}^\infty 1/(n\log n)$. Therefore the first series is divergent.

The denominator of the nth term of the second series is $3^{1+1/2+1/3+\cdots+1/n}$, and $1+1/2+1/3+\cdots+1/n$ grows like $\log n$. Hence the nth term of the second series is about

$$\frac{1}{3^{\log n}} = \frac{1}{n^{\log 3}}.$$

Now $\sum n^{-p}$ converges if $p > 1$ and $\log 3 > 1$, so the second series converges.

We shall give the details of this argument. Since

$$\sum_{k=1}^{n}\frac{1}{k} > \sum_{k=1}^{n}\int_k^{k+1}\frac{dt}{t} = \int_1^{n+1}\frac{dt}{t} = \log(n+1) > \log n,$$

we have

$$3^{1+1/2+1/3+\cdots+1/n} > 3^{\log n} = n^{\log 3}.$$

Hence the second series is dominated by $\sum n^{-\log 3}$. Since the latter converges, so does the second series.

Problem [11, 1951, A7]

Show that if the series $a_1+a_2+a_3+\cdots+a_n+\cdots$ converges, then the series $a_1+a_2/2+a_3/3+\cdots a_n/n+\cdots$ converges also.

Solution. This is a special case of a result sometimes called Abel's summation theorem: If the partial sums of the series $\sum a_n$ are bounded and the sequence $\{b_n\}$ decreases to zero, then $\sum a_n b_n$ is convergent.

[The text solution goes on to give a proof of Abel's summation theorem, which is substantially the same as the proof we have already given in Chapter 2.]

For the particular problem, take $b_n = 1/n$ and the result is immediate.

Problem [12, 1952, B5]

If the terms of a sequence $\{a_n\}$ are monotonic, and if $\sum_{n=1}^{\infty} a_n$ converges, show that $\sum_{n=1}^{\infty} n(a_n - a_{n+1})$ converges.

Solution 1. Since $\sum a_n$ converges, $\lim_{n\to\infty} a_n = 0$. Since the sequence is monotonic we have either

$$a_1 \geq a_2 \geq a_3 \geq \cdots \geq a_n \geq \cdots \geq 0,$$

or

$$a_1 \leq a_2 \leq a_3 \leq \cdots \leq a_n \leq \cdots \leq 0.$$

In the second case we can change the sign of each term and thus without loss of generality we need only consider the case where the a_n's are nonnegative and decrease to zero.

Let

$$S_k = \sum_{n=1}^{k} n(a_n - a_{n+1}).$$

Then

$$\begin{aligned} S_k &= a_1 + a_2(2-1) + \cdots + a_k(k-(k-1)) - k(a_{k+1}) \\ &= \sum_{n=1}^{k} a_n - ka_{k+1}. \end{aligned}$$

Now since $\sum_{n=1}^{\infty} a_n$ converges, the Cauchy criterion implies $\lim_{n\to\infty}(a_n + a_{n+1} + \cdots + a_{2n}) = 0$. But $a_n + a_{n+1} + \cdots + a_{2n} \geq na_{2n}$, and hence $\lim_{n\to\infty} 2na_{2n} = 0$; from this it follows that $\lim_{n\to\infty} na_{n+1} = 0$.

In the expression

$$S_k = \sum_{n=1}^{k} a_n - ka_{k+1}$$

both $\lim_{k\to\infty} \sum_{n=1}^{k} a_n$ and $\lim_{k\to\infty} ka_{k+1}$ exist, so

$$\lim_{k\to\infty} S_k = \lim_{k\to\infty} \sum_{n=1}^{k} a_n - \lim_{k\to\infty} (ka_{k+1}) = \lim_{k\to\infty} \sum_{n=1}^{k} a_n.$$

This establishes the desired convergence.

Solution 2. We may assume, as we have seen above, that the sequence $\{a_k\}$ decreases to zero. Then for each k,

$$a_k = \sum_{n=k}^{\infty} (a_n - a_{n+1}).$$

So

$$\begin{aligned}\sum_{k=1}^{\infty} a_k &= \sum_{k=1}^{\infty}\sum_{n=k}^{\infty} (a_n - a_{n+1}) = \sum_{n=1}^{\infty}\sum_{k=1}^{n} (a_n - a_{n+1}) \\ &= \sum_{n=1}^{\infty} n(a_n - a_{n+1}).\end{aligned}$$

Reversing the order of summation is justified because all terms are nonnegative.

Problem [13, 1953, B1]

Is the infinite series

$$\sum_{n=1}^{\infty} \frac{1}{n^{(n+1)/n}}$$

convergent? Prove your statement.

Solution. For every positive integer n, $n < 2^n$. Hence $n^{1/n} < 2$, so

$$\frac{1}{n^{(n+1)/n}} > \frac{1}{2n}.$$

Since $\sum_{n=1}^{\infty} \frac{1}{2n}$ diverges, so does

$$\sum_{n=1}^{\infty} \frac{1}{n^{(n+1)/n}}.$$

Problem [14, 1954, A6]

Suppose that $u_0, u_1, u_2, \ldots$ is a sequence of real numbers such that

$$u_n = \sum_{k=1}^{\infty} u_{n+k}^2 \quad \text{for } n = 0, 1, 2, \ldots .$$

Prove that if $\sum u_n$ converges, then $u_k = 0$ for all k.

Solution. It is obvious from the hypothesis that the terms u_n are nonincreasing and nonnegative. Thus, if any one term is zero, so are all its successors, and by induction using the hypothesis so are all its predecessors.

Suppose $\sum u_n$ converges. Let p be chosen so that $\sum_{n>p} u_n < 1$. Then, from the hypothesis,

$$u_p = \sum_{n>p} u_n^2 \leq \sum_{n>p} u_p u_n = u_p \sum_{n>p} u_n \leq u_p,$$

with equality at the last step only if $u_p = 0$. But, looking at the first member and the last member, we see that equality must hold throughout. Thus $u_p = 0$ and hence all the u's are zero, as we have shown above.

Problem [14, 1954, B2]

Assume as known the (true) fact that the alternating harmonic series

$$1 - 1/2 + 1/3 - 1/4 + 1/5 - 1/6 + 1/7 - 1/8 + \cdots$$

is convergent, and denote its sum by s. Rearrange the series as follows,

$$1 + 1/3 - 1/2 + 1/5 + 1/7 - 1/4 + 1/9 + 1/11 - 1/6 + \cdots .$$

Assume as known the (true) fact that this series is also convergent, and denote its sum by S. Denote by s_k, S_k the kth partial sum of the respective series. Prove the following statements.

1. $S_{3n} = s_{4n} + \frac{1}{2}s_{2n}$.
2. $S \neq s$.

Solution. First, we have

$$\begin{aligned} s_{4n} &= 1 - \frac{1}{2} + \frac{1}{3} - \frac{1}{4} + \frac{1}{5} - \frac{1}{6} + \cdots + \frac{1}{4n-1} - \frac{1}{4n}, \\ \frac{1}{2}s_{2n} &= \phantom{1 - {}} \frac{1}{2} \phantom{{}+\frac{1}{3}} - \frac{1}{4} \phantom{{}+\frac{1}{5}} + \frac{1}{6} + \cdots \phantom{{}+\frac{1}{4n-1}} - \frac{1}{4n}. \end{aligned}$$

Adding, we obtain

$$\begin{aligned} s_{4n} + \frac{1}{2}s_{2n} &= 1 + \frac{1}{3} - \frac{1}{2} + \frac{1}{5} + \frac{1}{7} - \frac{1}{4} + \cdots + \frac{1}{4n-3} + \frac{1}{4n-1} - \frac{1}{2n} \\ &= S_{3n}. \end{aligned}$$

For the second part of the conclusion, since the alternating harmonic series is an alternating series with terms that decrease to zero, $\lim_{n\to\infty} s_n = s$ exists. Moreover, $s > 1 - \frac{1}{2} = \frac{1}{2}$, so $s \neq 0$. Therefore

$$S = \lim_{n\to\infty} S_{3n} = \frac{3}{2}s \neq s.$$

Remark. Mathematics is full of arguments like this, that involve adding series term by term and cancelling some resulting zero terms. This can happen both with series of constants and series of functions. In this solution, the argument was carried out rigorously, using partial sums, but often enough it is shown for the infinite series themselves. There is something not quite trivial about this because it implicitly assumes that we can freely add or remove zero terms from an infinite sequence at will. Of course we want this to be true if our definition is reasonable, but since doing so can dramatically reindex the sequence of terms, it may not be immediately obvious that the sequence of partial sums will have the same limit behavior. Now, it is not hard to prove that such manipulations are always legal with our definition. It is worth noting that there are other less common notions of convergence for infinite series, and many of these lack this nice property.

Problem [15, 1954, A3]

Suppose that $\sum_{i=1}^{\infty} x_i$ is a convergent series of positive terms that monotonically decrease (that is, $x_1 \geq x_2 \geq x_3 \geq \cdots$). Let P denote the set of all numbers that are sums of some (finite or infinite) subseries of $\sum_{n=1}^{\infty} x_i$. Show that P is an interval if and only if

$$x_n \leq \sum_{i=n+1}^{\infty} x_i \text{ for every integer } n,$$

which we call condition (1).

Solution. Let $\mathbf{N}$ be the set of positive integers, and let J be a subset of $\mathbf{N}$. We write $S(J)$ for $\sum_{i \in J} x_i$. The problem requires us to show that the range of S is an interval if and only if (1) holds.

Suppose (1) fails for a given sequence. Let p be an index such that

$$x_p > \sum_{i>p} x_i.$$

Choose α so that $\sum_{i>p} x_i < \alpha < x_p$. Then there is no J for which $S(J) = \alpha$; for if $J \cap \{1, 2, \dots, p\} \neq \emptyset$, then $S(J) \geq x_p$ by the monotonicity of the x's, while if $J \cap \{1, 2, \dots, p\} = \emptyset$, then $S(J) \leq \sum_{i>p} x_i$. Since $\sum_{i>p} x_i$ and x_p are both in range(S), we see that range(S) is not an interval. Thus (1) is necessary in order that range(S) be an interval.

Now suppose (1) holds and $0 < y < S(\mathbf{N})$. We shall construct a set L such that $S(L) = y$. We define a sequence $n_1, n_2, n_3, \dots$ by induction as follows. Let n_1 be the least index for which

$$x_{n_1} < y.$$

(Such an index exists because $x_k \to 0$.) Assuming that $n_1, n_2, \dots, n_k$ have been chosen so that

$$x_{n_1} + x_{n_2} + \cdots + x_{n_k} < y,$$

let n_{k+1} be the least index exceeding n_k such that

$$x_{n_1} + x_{n_2} + \cdots + x_{n_k} + x_{n_{k+1}} < y.$$

(Again, such an index exists.)

Let $L = \{n_1, n_2, \dots\}$. Clearly

$$S(L) \le y,$$

which we call condition (2).

If $p \in N - L$, there is a least index k such that $n_k > p$. In choosing n_k we rejected p, hence

$$x_{n_1} + x_{n_2} + \cdots + x_{n_{k-1}} + x_p \ge y$$

and therefore

$$S(L) + x_p \ge y.$$

We split the remainder of the proof into two cases.

Case 1. (The set $N - L$ is finite.) Note that $N - L \ne \emptyset$, since in that case, $S(L) = S(\mathbf{N}) > y$, contradicting (2). Hence $N - L$ has a largest element that we can take to be p. Then

$$L = \{n_1, n_2, \dots, n_{k-1}\} \cup \{p+1, p+2, \dots\}.$$

Then combining previous results with (1),

$$S(L) = x_{n_1} + x_{n_2} + \cdots + x_{n_{k-1}} + \sum_{i>p} x_i \ge x_{n_1} + x_{n_2} + \cdots + x_{n_{k-1}} + x_p \ge y$$

which shows that $S(L) = y$.

Case 2. (The set $N - L$ is infinite.) Then for any $\epsilon > 0$, we can choose $p \in N - L$ so that $x_p < \epsilon$. Then we have

$$S(L) + \epsilon > S(L) + x_p \ge y.$$

Since ϵ is arbitrary, we obtain $S(L) \ge y$. So again we must have $S(L) = y$.

Since $S(\emptyset) = 0$, it follows that range$(S) = [0, S(\mathbf{N})]$. Thus (1) is both necessary and sufficient in order that range(S) be an interval.

Problem [16, 1956, B6]

Given $T_1 = 2$, $T_{n+1} = T_n^2 - T_n + 1$, $n > 0$, prove:

1. If $m \ne n$, T_m and T_n have no common factor greater than 1.
2. $\sum_{i=1}^{\infty}(1/T_i) = 1$.

Solution. The first few members of the sequence are $T_1 = 2, T_2 = 3, T_3 = 7$. We shall prove by induction that

$$T_{n+1} = 1 + \prod_{i=1}^{n} T_i \quad \text{for } n \geq 1.$$

This is true for $n = 1$. Suppose it is true for $n = k$. Then

$$\begin{aligned} T_{k+2} &= 1 + T_{k+1}(T_{k+1} - 1) \\ &= 1 + T_{k+1}\left[\prod_{i=1}^{k} T_i\right] \\ &= 1 + \prod_{i=1}^{k+1} T_i. \end{aligned}$$

(The first step is by the given recursion, the second by the inductive hypthesis.) This completes the inductive proof.

Now suppose $m \neq n$, say $m < n$. Then our product representation shows that T_m divides $T_n - 1$, so T_m and T_n are relatively prime. This is the first statement we were to prove.

Next we prove by induction that

$$\sum_{i=1}^{n} \frac{1}{T_i} = 1 - \frac{1}{T_{n+1} - 1}$$

for all n. This is true for $n = 1$ and, if it is true for $n = k$, we have

$$\begin{aligned} \sum_{i=1}^{k+1} \frac{1}{T_i} &= 1 - \frac{1}{T_{k+1} - 1} + \frac{1}{T_{k+1}} \\ &= 1 - \frac{1}{T_{k+1}(T_{k+1} - 1)} \\ &= 1 - \frac{1}{T_{k+2} - 1}, \end{aligned}$$

completing the induction.

Since $T_n \to \infty$ as $n \to \infty$, it follows from the above that

$$\sum_{i=1}^{\infty} \frac{1}{T_i} = 1,$$

as required.

Problem [18, 1958, A6]

What is the smallest amount that may be invested at interest rate i, compounded annually, in order that one may withdraw 1 dollar at the end of the first year, 4 dollars at the end of the second year, ... , n^2 dollars at the end of the nth year, in perpetuity.

Answer: $(1+i)(2+i)/i^3$.

Solution. The present value of one dollar to be paid after n years is $(1+i)^{-n}$ dollars. Hence the value in dollars of the given annuity is

$$\sum_{n=1}^{\infty} n^2(1+i)^{-n}.$$

Since

$$\frac{1}{1-x} = \sum_{n=0}^{\infty} x^n,$$

we have

$$\frac{x}{(1-x)^2} = x\frac{d}{dx}\left(\frac{1}{1-x}\right) = \sum_{n=1}^{\infty} nx^n$$

and

$$\frac{x+x^2}{(1-x)^3} = x\frac{d}{dx}\left(\frac{x}{(1-x)^2}\right) = \sum_{n=1}^{\infty} n^2x^n,$$

for $|x| < 1$. Putting $x = 1/(1+i)$, we obtain

$$\sum_{n=1}^{\infty} n^2(1+i)^{-n} = \frac{(1+i)(2+i)}{i^3}.$$

(At 6 percent interest, the cost of the annuity would be \$10,109.26.)

Problem [19, 1958, B5]

The lengths of successive segments of a broken line are represented by the successive terms of the harmonic progression $1, 1/2, 1/3, \ldots, 1/n, \ldots$. Each segment makes with the preceding segment a given angle θ. What is the distance and what is the direction of the limiting point (if there is one) from the initial point of the first segment?

Answer: The limiting point is at a distance

$$\sqrt{\left(\log\left(2\sin\frac{\theta}{2}\right)\right)^2 + \frac{1}{4}(\theta-\pi)^2}$$

and at an angle

$$\pi - \theta + \arg\left[\log\left(2\sin\frac{\theta}{2}\right) + \frac{i}{2}(\theta - \pi)\right].$$

Solution 1. We may identify the plane with the complex number plane in such a way that the first segment extends from 0 to 1. Then the next segment extends from 1 to $1 + \frac{1}{2}e^{i\theta}$, since this segment represents the complex number $\frac{1}{2}e^{i\theta}$.

The nth segment of the path represents the complex number $(1/n)e^{i(n-1)\theta}$, and when added to the previous $(n-1)$ segments, the sum ends at the point $\sum_{p=1}^{n}(1/p)e^{i(p-1)\theta}$. Thus the question really concerns the convergence and evaluation of

$$\sum_{p=1}^{\infty}(1/p)e^{i(p-1)\theta}.$$

This becomes the harmonic series if $\theta = 0$ (or a multiple of 2π) and does not converge in this case. We shall show that the series converges in all other cases.

[Again the official solution now refers to Abel's Theorem, and we omit the proof and discussion, since it is superceded by the treatment in Chapter 2.]

For the present case, take $b_p = 1/p$ and $a_p = e^{i(p-1)\theta}$. Since the a_p's form a geometric progression, and since we are assuming $e^{i\theta} \neq 1$, we have

$$\left|\sum_{p=1}^{n} a_p\right| = \left|\frac{1-e^{in\theta}}{1-e^{i\theta}}\right| \leq \frac{2}{|1-e^{i\theta}|}.$$

Thus the partial sums are bounded, Abel's Theorem applies, and the series in question converges whenever $e^{i\theta} \neq 1$.

To evaluate the series, we can use another theorem of Abel. If $\sum_{p=1}^{\infty} c_p$ converges, then

$$\lim_{r\to 1^-}\sum_{p=1}^{\infty} c_p r^p = \sum_{p=1}^{\infty} c_p$$

where the limit is taken for r increasing to 1 through real values.

In the present case,

$$\sum \frac{1}{p}e^{i(p-1)\theta} = e^{-i\theta}\lim_{r\to 1^-}\sum\frac{1}{p}(re^{i\theta})^p.$$

Putting $z = re^{i\theta}$, we recognize the series on the right as the Taylor's series expansion for the principal value of $-\log(1-z)$, valid for $|z| < 1$. Hence

$$\sum \frac{1}{p}e^{i(p-1)\theta} = -e^{-i\theta}\lim_{r\to 1^-}\log(1 - re^{i\theta}) = -e^{-i\theta}\log(1 - e^{i\theta}).$$

Now

$$1 - e^{i\theta} = 2\left(\sin\frac{1}{2}\theta\right)e^{(1/2)i(\theta-\pi)} \quad \text{for } 0 < \theta < 2\pi$$

and here $2\sin\frac{1}{2}\theta > 0$ and $\frac{1}{2}(\theta-\pi)$ is the principal value of the argument (since $-\pi < \frac{1}{2}(\theta-\pi) < \pi$). Therefore

$$\log(1-e^{i\theta}) = \log\left(2\sin\frac{1}{2}\theta\right) + \frac{1}{2}i(\theta-\pi)$$

and

$$\sum \frac{1}{p}e^{i(p-1)\theta} = -e^{-i\theta}\left\{\log\left(2\sin\frac{1}{2}\theta\right) + \frac{1}{2}i(\theta-\pi)\right\}.$$

The stated distance and angle now follow immediately.

Solution 2. One can prove the relation

$$\sum_{p=1}^{\infty} \frac{1}{p}e^{ip\theta} = -\log(1-e^{i\theta}) = -\log\left(2\sin\frac{1}{2}\theta\right) + \frac{1}{2}i(\pi-\theta)$$

using the theory of Fourier series. The Fourier series of $\frac{1}{2}(\pi-\theta)$ on the interval $[0, 2\pi]$ is readily found to be

$$\sum_{p=1}^{\infty} \frac{1}{p}\sin p\theta.$$

With slightly more difficulty the Fourier series of $\log(2\sin\frac{1}{2}\theta)$ is found to be

$$-\sum_{p=1}^{\infty} \frac{1}{p}\cos p\theta.$$

Since both functions in the claimed representation for $\sum \frac{1}{p}e^{ip\theta}$ have continuous derivatives on the open interval $(0, 2\pi)$, they are represented by their Fourier series on this interval, i.e., the series converges and

$$\sum_{p=1}^{\infty} \frac{1}{p}\cos p\theta = -\log\left(2\sin\frac{1}{2}\theta\right),$$

$$\sum_{p=1}^{\infty} \frac{1}{p}\sin p\theta = \frac{1}{2}(\pi-\theta), \quad \text{for } 0 < \theta < \pi.$$

Combining these relations and using $e^{ip\theta} = \cos p\theta + i\sin p\theta$ we obtain the assertion.

Problem [20, 1959, B2]

Let c be a positive real number. Prove that c can be expressed in infinitely many ways as a sum of infinitely many distinct terms selected from the sequence

$$1/10, 1/20, \ldots, 1/10n, \ldots.$$

Solution. [This problem is a special case of a more general result that we proved in Chapter 3. The official solution to this problem essentially follows our proof of that result.]

Problem [21, 1960, B2]

Evaluate the double series

$$\sum_{j=0}^{\infty}\sum_{k=0}^{\infty} 2^{-3k-j-(k+j)^2}.$$

Answer: The sum is $4/3$.

Solution. If we write out the values of the double sequence of terms, it becomes evident that each of the terms 2^{-2m} occurs exactly once. Since all the terms are positive, the double series may be rearranged to make the simple series

$$\sum_{m=0}^{\infty} 2^{-2m}$$

which sums to $4/3$. The sum of the original double series is therefore $4/3$. To formalize this argument one must prove that $(j,k) \mapsto (k+j)^2 + 3k + j$ is a bijection from the set of ordered pairs of nonnegative integers to the set of nonnegative even integers. This is straightforward but long.

A direct analytic argument can be given as follows: Sum the double series along the diagonals defined by $j + k = n$, and then sum on n. We find

$$\begin{aligned}
\sum_{j=0}^{\infty}\sum_{k=0}^{\infty} 2^{-3k-j-(k+j)^2} &= \sum_{n=0}^{\infty}\sum_{k=0}^{n} 2^{-2k-n-n^2} \\
&= \sum_{n=0}^{\infty}\left[\frac{4}{3}(1-2^{-2n-2})2^{-n-n^2}\right] \\
&= \frac{4}{3}\left[\sum_{n=0}^{\infty} 2^{-n-n^2} - \sum_{n=0}^{\infty} 2^{-(n+1)(n+2)}\right] \\
&= \frac{4}{3}\left[\sum_{n=0}^{\infty} 2^{-n(n+1)} - \sum_{m=1}^{\infty} 2^{-m(m+1)}\right] \\
&= \frac{4}{3}.
\end{aligned}$$

At the third step, the separation of the sum into two sums is permissible because the new sums are convergent.

Problem [21, 1960, B6]

Any positive integer may be written in the form $n = 2^k(2l+1)$. Let $a_n = e^{-k}$ and $b_n = a_1 a_2 a_3 \cdots a_n$. Prove that $\sum b_n$ converges.

Solution. It is clear that $a_n = e^0 = 1$ if n is odd and $a_n \le e^{-1}$ if n is even. Therefore

$$b_{2k} = a_1 a_2 \cdots a_{2k} \le e^{-k},$$

and

$$b_{2k+1} \le e^{-k}.$$

Therefore,

$$\begin{aligned} b_1 + b_2 + \cdots + b_{2k} &< b_1 + b_2 + \cdots + b_{2k+1} \\ &\le 1 + 2e^{-1} + 2e^{-2} + \cdots + 2e^{-2k} < 1 + \frac{2e^{-1}}{1 - e^{-1}}. \end{aligned}$$

Thus the partial sums of $\sum b_n$ are bounded. Since the series has positive terms, it converges.

Problem [24, 1963, B5]

Let $\{a_n\}$ be a sequence of real numbers satisfying the inequalities

$$0 \le a_k \le 100 a_n \quad \text{for } n \le k \le 2n \quad \text{and} \quad n = 1, 2, \ldots,$$

and such that the series

$$\sum_{n=0}^{\infty} a_n$$

converges. Prove that

$$\lim_{n \to \infty} n a_n = 0.$$

Solution. For each positive integer k let

$$S_k = \sum_{n \ge k/2}^{k} a_n.$$

Since $\sum a_n$ converges, we have $S_k \to 0$ as $k \to \infty$.

Rewriting the given condition slightly we have

$$0 \le a_k \le 100 a_n \quad \text{for } \frac{1}{2}k \le n \le k.$$

For each k there are at least $k/2$ integers n satisfying the double inequality. Adding these inequalities we have, therefore,

$$\frac{1}{2} k a_k \le 100 S_k.$$

Hence

$$\limsup k a_k \leq 200 \lim S_k = 0.$$

Since $a_k \geq 0$, we conclude $\lim k a_k = 0$.

Problem [25, 1964, A3]

Let $P_1, P_2, \dots$ be a sequence of distinct points that is dense in the interval $(0, 1)$. The points $P_1, P_2, \dots, P_{n-1}$ decompose the interval into n parts, and P_n decomposes one of these into two parts. Let a_n and b_n be the lengths of these two intervals. Prove that

$$\sum_{n=1}^{\infty} a_n b_n (a_n + b_n) = \frac{1}{3}.$$

(A sequence of points in an interval is said to be dense when every subinterval contains at least one point of the sequence.)

Solution. Let S_n be the sum of the cubes of the lengths of the segments formed by the points $P_1, P_2, \dots, P_{n-1}, P_n$. Take $S_0 = 1$. We can obtain S_n from S_{n-1} by removing the term $(a_n + b_n)^3$ and replacing it by $a_n^3 + b_n^3$, leaving all other terms fixed.

Hence

$$\begin{aligned} S_{n-1} - S_n &= (a_n + b_n)^3 - a_n^3 - b_n^3 \\ &= 3a_n b_n (a_n + b_n). \end{aligned}$$

Therefore

$$1 - S_k = S_0 - S_k = \sum_{n=1}^{k} 3a_n b_n (a_n + b_n).$$

Since, as we shall prove below, $\lim_{k\to\infty} S_k = 0$,

$$\sum_{n=1}^{\infty} 3a_n b_n (a_n + b_n) = 1,$$

and the required result follows immediately.

We now show that $\lim_{k\to\infty} S_k = 0$. Let t be a positive integer. Because the set $\{P_i \mid i = 1, 2, 3, \dots\}$ is dense, we can choose an integer q so large that $\{P_1, P_2, \dots, P_q\}$ meets each of the intervals

$$[0, 1/t], [1/t, 2/t], \dots, [(t-1)/t, 1].$$

Suppose $k \geq q$ and let $l_0, l_1, \dots, l_k$ be the lengths of the intervals determined by the division points $P_1, P_2, \dots, P_k$. Then each of these intervals has length at most $2/t$, so

$$S_k = \sum_{i=0}^{k} l_i^3 \le \left(\frac{2}{t}\right)^2 \sum_{i=0}^{k} l_i = \frac{4}{t^2}.$$

Since t was arbitrary, this proves

$$\lim S_k = 0.$$

Problem [25, 1964, A5]

Prove that there is a constant K such that the following inequality holds for any sequence of positive numbers $a_1, a_2, a_3, \dots$:

$$\sum_{n=1}^{\infty} \frac{n}{a_1 + a_2 + \cdots + a_n} \le K \sum_{n=1}^{\infty} \frac{1}{a_n}.$$

Solution. Let $k = 2t$ be some fixed even positive integer. We shall prove

$$\sum_{n=1}^{k} \frac{n}{a_1 + a_2 + \cdots + a_n} \le 4 \sum_{n=1}^{k} \frac{1}{a_n}.$$

From this inequality, the required inequality for infinite sums, with $K = 4$, follows immediately.

Let $b_1, b_2, \dots, b_k$ be the terms $a_1, a_2, \dots, a_k$ enumerated in increasing order. For $1 \le p \le t$ we have

$$\begin{aligned} a_1 + a_2 + \cdots + a_{2p} &\ge a_1 + a_2 + \cdots + a_{2p-1} \\ &\ge b_1 + b_2 + \cdots + b_{2p-1} \ge p b_p, \end{aligned}$$

since all terms are positive and since the last p terms are at least b_p. Therefore

$$\frac{2p-1}{a_1 + a_2 + \cdots + a_{2p-1}} \le \frac{2p-1}{p b_p} < \frac{2}{b_p}.$$

Also

$$\frac{2p}{a_1 + a_2 + \cdots + a_{2p}} \le \frac{2}{b_p}.$$

Hence

$$\frac{2p-1}{a_1 + a_2 + \cdots + a_{2p-1}} + \frac{2p}{a_1 + a_2 + \cdots + a_{2p}} < \frac{4}{b_p}.$$

Thus

$$\sum_{p=1}^{t} \left[\frac{2p-1}{a_1 + a_2 + \cdots + a_{2p-1}} + \frac{2p}{a_1 + a_2 + \cdots + a_{2p}}\right] < \sum_{p=1}^{t} \frac{4}{b_p}.$$

Rewriting the left-hand sum, we have

$$\sum_{n=1}^{k} \frac{n}{a_1 + a_2 + \cdots + a_n} \leq \sum_{p=1}^{t} \frac{4}{b_p} < \sum_{p=1}^{k} \frac{4}{b_p} = 4 \sum_{n=1}^{k} \frac{1}{a_n}.$$

If the series $\sum_{n=1}^{\infty}(1/a_n)$ diverges, the given inequality is deemed to be satisfied automatically. Otherwise we have

$$\sum_{n=1}^{k} \frac{n}{a_1 + a_2 + \cdots + a_n} \leq 4 \sum_{n=1}^{\infty} \frac{1}{a_n}$$

and hence

$$\sum_{n=1}^{\infty} \frac{n}{a_1 + a_2 + \cdots + a_n} \leq 4 \sum_{n=1}^{\infty} \frac{1}{a_n}.$$

Problem [25, 1964, B1]

Let u_k $(k = 1, 2, \ldots)$ be a sequence of integers, and let V_n be the number of those that are less than or equal to n. Show that if

$$\sum_{k=1}^{\infty} 1/u_k < \infty,$$

then

$$\lim_{n \to \infty} V_n/n = 0.$$

Solution. It must have been intended that the u's be positive, since otherwise we could have $\sum 1/u_k$ convergent but all the V's infinite. Hence we assume that $\sum 1/u_k$ is a convergent series of positive terms. Then $u_k \to \infty$, and the V's are finite. Since the convergence of a series of positive terms is not affected by rearrangement, we can assume that $u_1 \leq u_2 \leq u_3 \cdots$. Then, for a fixed p and any n so large that $V_n > p$ we have

$$\sum_{k=p+1}^{V_n} 1/u_k \geq \frac{V_n - p}{n},$$

because there are $V_n - p$ terms in the sum and each is at least $1/n$.

Therefore

$$\limsup_{n \to \infty} \frac{V_n}{n} \leq \sum_{k=p+1}^{\infty} 1/u_k.$$

Since this is true for any p we have

$$\limsup_{n \to \infty} \frac{V_n}{n} \leq \lim_{p \to \infty} \sum_{k=p+1}^{\infty} 1/u_k = 0.$$

But $V_n/n \geq 0$ for all n and therefore $\lim_{n \to \infty} V_n/n = 0$.

Problem [25, 1964, B5]

Let u_n ($n = 1, 2, 3, \ldots$) denote the least common multiple of the first n terms of a strictly increasing sequence of positive integers (for example, the sequence 1, 2, 3, 4, 5, 6, 10, 12). Prove that the series

$$\sum_{n=1}^{\infty} 1/u_n$$

is convergent.

Solution 1. Let u be a positive integer. For each divisor d of u exceeding $\sqrt{u}$ there is another u/d that is less than $\sqrt{u}$. Hence at least half of the positive divisors of u are less than $\sqrt{u}$, so the number of positive divisors is at most $2\sqrt{u}$.

Now u_n, being the least common multiple of n distinct positive integers, has at least n positive divisors, so $2\sqrt{u_n} \geq n$. Therefore, $\sum 1/u_n$ is dominated by the convergent series $\sum 4/n^2$ and is itself convergent.

Solution 2 [Due to R.L. Graham]. The following proof shows that the upper bound for such sums is 2.

Let $a_1, a_2, \ldots$ be a strictly increasing sequence of positive integers. Then

$$\begin{aligned}\frac{1}{a_1} &= \sum_{n=2}^{\infty}\left(\frac{1}{a_{n-1}} - \frac{1}{a_n}\right) = \sum_{n=2}^{\infty} \frac{a_n - a_{n-1}}{a_{n-1}a_n}\\ &\geq \sum_{n=2}^{\infty} \frac{\text{g.c.d.}\{a_{n-1}, a_n\}}{a_{n-1}a_n} = \sum_{n=2}^{\infty} \frac{1}{\text{l.c.m.}\{a_{n-1}, a_n\}}\\ &\geq \sum_{n=2}^{\infty} \frac{1}{\text{l.c.m.}\{a_1, a_2, \ldots, a_n\}} = \sum_{n=2}^{\infty} \frac{1}{u_n}.\end{aligned}$$

(Here g.c.d. stands for greatest common divisor and l.c.m. stands for least common multiple.) Since $u_1 = a_1$, we have

$$\sum_{n=1}^{\infty} \frac{1}{u_n} \leq \frac{2}{a_1}.$$

There are many a-sequences for which $\sum 1/u_n = 2$; for example, $1, 2, 4, 8, \ldots$ or $1, 2, 3, 6, 12, \ldots, 3 \cdot 2^{n-3}, \ldots$.

Problem [27, 1966, B3]

Show that if the series

$$\sum_{n=1}^{\infty} \frac{1}{p_n}$$

is convergent, where $p_1, p_2, p_3, \ldots, p_n, \ldots$ are positive real numbers, then the series

$$\sum_{n=1}^{\infty} \frac{n^2}{(p_1 + p_2 + \cdots + p_n)^2} p_n$$

is also convergent.

Solution. Set $q_n = p_1 + p_2 + \cdots + p_n$, with $q_0 = 0$. We are led to estimate $S_N = \sum_{n=1}^{N} (n/q_n)^2 (q_n - q_{n-1})$ in terms of $T = \sum_{n=1}^{\infty} 1/p_n$. Note that

$$\begin{aligned} S_N &\le \frac{1}{p_1} + \sum_{n=2}^{N} \frac{n^2}{q_n q_{n-1}} (q_n - q_{n-1}) \\ &= \frac{1}{p_1} + \sum_{n=2}^{N} \frac{n^2}{q_{n-1}} - \sum_{n=2}^{N} \frac{n^2}{q_n} \\ &= \frac{1}{p_1} + \sum_{n=1}^{N-1} \frac{(n+1)^2}{q_n} - \sum_{n=2}^{N} \frac{n^2}{q_n} \\ &\le \frac{5}{p_1} + 2\sum_{n=1}^{N} \frac{n}{q_n} + \sum_{n=2}^{N} \frac{1}{q_n}. \end{aligned}$$

By Schwarz's inequality,

$$\left(\sum_{n=2}^{N} \frac{n}{q_n} \right)^2 \le \sum_{n=2}^{N} \frac{n^2}{q_n^2} p_n \sum_{n=2}^{N} \frac{1}{p_n}$$

and thus

$$S_N \le \frac{5}{p_1} + 2\sqrt{S_N T} + T.$$

This inequality is quadratic in $\sqrt{S_N}$ and by the quadratic inequality we have

$$\sqrt{S_N} \le \sqrt{T} + \sqrt{2T + 5/p_1}.$$

Thus our series is positive and has bounded partial sums, hence it must converge.

Problem [30,1969,B5]

Let $a_1 < a_2 < a_3 < \cdots$ be an increasing sequence of positive integers. Let the series $\sum_{n=1}^{\infty} 1/a_n$ be convergent. For any number x let $k(x)$ be the number of the a_n's that do not exceed x. Show that $\lim_{x\to\infty} k(x)/x = 0$.

Solution. The following proof shows that it is not necessary to stipulate that the a_n be integers. Suppose that for some $\epsilon > 0$ there are $x_j \to \infty$ with $k(x_j)/x_j \ge \epsilon$. Note that if $1 \le n \le k(x_j)$, then (because the a_n increase) $a_n \le a_{k(x_j)} \le x_j$ and $1/a_n \ge 1/x_j$. Now for any positive integer N,

$$\sum_{n=N}^{\infty} 1/a_n \geq \sup_j \sum_{n=N}^{k(x_j)} 1/a_n \geq \sup_j \frac{k(x_j) - N}{x_j} \geq \sup_j (\epsilon - N/x_j) = \epsilon.$$

But this contradicts the convergence of $\sum_{n=1}^{\infty} 1/a_n$, which implies

$$\lim_{N \to \infty} \sum_{n=N}^{\infty} 1/a_n = 0.$$

Problem [36,1975,B5]

Let $f_0(x) = e^x$ and $f_{n+1}(x) = x f_n'(x)$ for $n = 0, 1, 2, \ldots$. Show that

$$\sum_{n=0}^{\infty} \frac{f_n(1)}{n!} = e^e.$$

Solution. Since $f_0(x) = \sum_{k=0}^{\infty} x^k/k!$, a straightforward mathematical induction proves that $f_n(x) = \sum_{k=0}^{\infty} k^n x^k / k!$. Then, since all terms are positive, one has

$$\sum_{n=0}^{\infty} \frac{f_n(1)}{n!} = \sum_{n=0}^{\infty} \sum_{k=0}^{\infty} \frac{k^n}{k!n!} = \sum_{k=0}^{\infty} \frac{1}{k!} \sum_{n=0}^{\infty} \frac{k^n}{n!} = \sum_{k=0}^{\infty} \frac{e^k}{k!} = e^e.$$

Problem [39, 1978, B2]

Express

$$\sum_{m=1}^{\infty} \sum_{n=1}^{\infty} \frac{1}{m^2 n + m n^2 + 2mn}$$

as a rational number.

Answer: $7/4$.

Solution. Let S be the desired sum. Then

$$\begin{aligned}
S &= \sum_{n=1}^{\infty} \frac{1}{n} \sum_{m=1}^{\infty} \frac{1}{n+2} \left(\frac{1}{m} - \frac{1}{m+n+2} \right) \\
&= \sum_{n=1}^{\infty} \frac{1}{n(n+2)} \left[\left(1 - \frac{1}{n+3} \right) + \left(\frac{1}{2} - \frac{1}{n+4} \right) + \left(\frac{1}{3} - \frac{1}{n+5} \right) + \cdots \right] \\
&= \frac{1}{2} \sum_{n=1}^{\infty} \left(\frac{1}{n} - \frac{1}{n+2} \right) \left[\left(1 - \frac{1}{n+3} \right) + \left(\frac{1}{2} - \frac{1}{n+4} \right) + \cdots \right]
\end{aligned}$$

Notice that the bracketed expression is in fact a telescoping sum for each fixed n. Hence

$$
\begin{aligned}
2S &= \sum_{n=1}^{\infty}\left(\frac{1}{n}-\frac{1}{n+2}\right) \\
&\quad \times \lim_{k\to\infty}\left[1+\frac{1}{2}+\cdots+\frac{1}{n+2}-\frac{1}{k}-\frac{1}{k+1}-\cdots-\frac{1}{k+n+1}\right] \\
&= \sum_{n=1}^{\infty}\left(\frac{1}{n}-\frac{1}{n+2}\right)\left(1+\frac{1}{2}+\cdots+\frac{1}{n+2}\right) \\
&= \lim_{h\to\infty}\left[\left(1-\frac{1}{3}\right)\left(1+\frac{1}{2}+\frac{1}{3}\right)+\left(\frac{1}{2}-\frac{1}{4}\right)\left(1+\frac{1}{2}+\frac{1}{3}+\frac{1}{4}\right)\right. \\
&\quad \left.+\cdots+\left(\frac{1}{h}-\frac{1}{h+2}\right)\left(1+\frac{1}{2}+\cdots+\frac{1}{h}\right)\right] \\
&= \lim_{h\to\infty}\left[1\left(1+\frac{1}{2}+\frac{1}{3}\right)+\frac{1}{2}\left(1+\frac{1}{2}+\frac{1}{3}+\frac{1}{4}\right)+\frac{1}{3}\left(\frac{1}{4}+\frac{1}{5}\right)\right. \\
&\quad +\frac{1}{4}\left(\frac{1}{5}+\frac{1}{6}\right)+\cdots+\frac{1}{h}\left(\frac{1}{h+1}+\frac{1}{h+2}\right) \\
&\quad \left.-\frac{1}{h+1}\left(1+\frac{1}{2}+\cdots+\frac{1}{h-1}\right)-\frac{1}{h+2}\left(1+\frac{1}{2}+\cdots+\frac{1}{h}\right)\right] \\
&= \frac{6+3+2}{6}+\frac{12+6+4+3}{2\cdot 12}+\sum_{n=3}^{\infty}\frac{1}{n(n+1)}+\sum_{n=3}^{\infty}\frac{1}{n(n+2)} \\
&= \frac{11}{6}+\frac{25}{24}+\frac{1}{3}+\frac{1}{6}+\frac{1}{8} \\
&= \frac{7}{2}
\end{aligned}
$$

where we have identified the last two series as known telescoping series.

Thus $S = 7/4$.

Problem [43, 1982, A2]

For positive real x, let

$$B_n(x) = 1^x + 2^x + 3^x + \cdots + n^x.$$

Prove or disprove the convergence of

$$\sum_{n=2}^{\infty}\frac{B_n(\log_n 2)}{(n\log_2 n)^2}.$$

Answer: The series is convergent.

Solution. Since $x = \log_2 n > 0$, $B_n(x) = 1^x + 2^x + \cdots + n^x \le n(n^x)$ and

$$0 \le \frac{B_n(\log_n 2)}{(n \log_2 n)^2} \le \frac{n \cdot n^{\log_n 2}}{(n \log_2 n)^2} = \frac{2}{n(\log_2 n)^2}.$$

Notice finally that the series whose nth term is given by the expression at the right hand of this chain of inequalities converges by the Integral Test or by the Abel-Dini scale, so the series converges by the Comparison Test.

Problem [45, 1984, A2]

Express $\sum_{k=1}^{\infty}(6^k/(3^{k+1} - 2^{k+1})(3^k - 2^k))$ as a rational number.

Answer: The sum is 2.

Solution. Let $S(n)$ denote the nth partial sum of the given series. Then

$$S(n) = \sum_{k=1}^{n}\left[\frac{3^k}{3^k - 2^k} - \frac{3^{k+1}}{3^{k+1} - 2^{k+1}}\right] = 3 - \frac{3^{n+1}}{3^{n+1} - 2^{n+1}},$$

and the series converges to $\lim_{n\to\infty} S(n) = 2$.

Problem [47, 1986, A3]

Evaluate $\sum_{n=0}^{\infty}$ Arccot $(n^2 + n + 1)$, where Arccot t for $t \ge 0$ denotes the number θ in the interval $0 < \theta \le \pi/2$ with $\cot\theta = t$.

Answer: The series converges to $\pi/2$.

Solution 1. If $\alpha = \operatorname{Arccot} x$ and $\beta = \operatorname{Arccot} y$ for some some $x, y > 0$, then the addition formula

$$\cot(\alpha + \beta) = \frac{\cot\alpha \cot\beta - 1}{\cot\alpha + \cot\beta}$$

shows that

$$\operatorname{Arccot} x + \operatorname{Arccot} y = \operatorname{Arccot} \frac{xy - 1}{x + y}$$

provided that $\operatorname{Arccot} x + \operatorname{Arccot} y \le \pi/2$. The latter condition is equivalent to $\operatorname{Arccot} x \le \operatorname{Arccot}(1/y)$, which is equivalent to $x \ge 1/y$, and hence equivalent to $xy \ge 1$. Verifying the $xy \ge 1$ condition at each step, we use the previous equation to compute the first few partial sums,

$$\operatorname{Arccot} 1 = \operatorname{Arccot} 1,$$

$$\operatorname{Arccot} 1 + \operatorname{Arccot} 3 = \operatorname{Arccot}(1/2),$$

$$\operatorname{Arccot} 1 + \operatorname{Arccot} 3 + \operatorname{Arccot} 7 = \operatorname{Arccot}(1/3),$$

and guess that $\sum_{n=0}^{m-1} \text{Arccot}\,(n^2+n+1) = \text{Arccot}\,(1/m)$ for all $m \geq 1$. This is easily proved by induction on m: the base case is above, and the inductive step is

$$\begin{aligned}\sum_{n=0}^{m} \text{Arccot}\,(n^2+n+1) &= \text{Arccot}\,(m^2+m+1) + \sum_{n=0}^{m-1} \text{Arccot}\,(n^2+n+1) \\ &= \text{Arccot}\,(m^2+m+1) + \text{Arccot}\left(\frac{1}{m}\right) \\ &= \text{Arccot}\left(\frac{(m^2+m+1)/m - 1}{m^2+m+1+1/m}\right) \\ &= \text{Arccot}\left(\frac{1}{m+1}\right).\end{aligned}$$

.

Hence

$$\sum_{n=0}^{\infty} \text{Arccot}\,(n^2+n+1) = \lim_{m\to\infty} \text{Arccot}\left(\frac{1}{m}\right) = \text{Arccot}\,(0) = \frac{\pi}{2}.$$

Solution 2. For real $a \geq 0$ and $b \neq 0$, Arccot (a/b), is the argument (between $-\pi/2$ and $\pi/2$) of the complex number $a + bi$. Therefore, if any three complex numbers satisfy $(a+bi)(c+di) = (e+fi)$, where $a, c, e \geq 0$ and $b, d, f \neq 0$, then Arccot $(a/b) +$ Arccot $(c/d) =$ Arccot (e/f). Factoring the polynomial $n^2+n+1+i$ yields

$$(n^2+n+1+i) = (n+i)(n+1-i).$$

Taking arguments, we find that Arccot $(n^2+n+1) =$ Arccot $n -$ Arccot $(n+1)$. (This identity can also be proved from the difference formula for the cotangent, but then one needs to guess the identity in advance.) The series $\sum_{n=0}^{\infty} \text{Arccot}\,(n^2+n+1)$ telescopes to $\lim_{n\to\infty}(\text{Arccot}\,(0) - \text{Arccot}\,(n+1)) = \pi/2$.

Related question: Evaluate the infinite series

$$\sum_{n=1}^{\infty} \text{Arctan}\left(\frac{2}{n^2}\right), \quad \sum_{n=1}^{\infty} \text{Arctan}\left(\frac{8n}{n^4-2n^2+5}\right).$$

Problem [48, 1987, A6]

For each positive integer n, let $a(n)$ be the number of zeroes in the base 3 representation of n. For which positive real numbers x does the series

$$\sum_{n=1}^{\infty} \frac{x^{a(n)}}{n^3}$$

converge?

Answer: For positive real x, the series converges if and only if $x < 25$.

Solution. The integer $n \geq 1$ has exactly $k+1$ digits in base 3 if and only if $3^k \leq n < 3^{k+1}$. Define

$$S_k = \sum_{n=3^k}^{3^{k+1}-1} \frac{x^{a(n)}}{n^3}, \quad \text{and} \quad T_k = \sum_{n=3^k}^{3^{k+1}-1} x^{a(n)}.$$

The given series $\sum_{n=1}^{\infty} x^{a(n)}/n^3$ has all terms positive, so it will converge if and only if $\sum_{k=0}^{\infty} S_k$ converges. For $3^k \leq n < 3^{k+1}$, we have $3^{3k} \leq n^3 < 3^{3k+3}$, so $T_k/3^{3k+3} \leq S_k \leq T_k/3^{3k}$. Therefore $\sum_{k=0}^{\infty} S_k$ converges if and only if $\sum_{k=0}^{\infty} T_k/3^{3k}$ converges. The number of n with $k+1$ digits base 3 and satisfying $a(n) = i$ is $\binom{k}{i}2^{k+1-i}$, because there are $\binom{k}{i}$ possibilities for the set of positions of the i zero digits (since the leading digit cannot be zero), and then 2^{k+1-i} ways to select 1 or 2 as each of the remaining digits. Therefore

$$T_k = \sum_{i=0}^{k} \binom{k}{i} 2^{k+1-i} x^i = (x+2)^k.$$

Hence

$$\sum_{k=0}^{\infty} T_k/3^{3k} = \sum_{k=0}^{\infty} \left(\frac{x+2}{27}\right)^k,$$

which converges if and only if $|(x+2)/27| < 1$. For positive x, this condition is equivalent to $0 < x < 25$.

Remark (by Kedlaya, Poonen, Vakil). More generally, let $\alpha_k(n)$ be the number of zeros in the base k expansion of n, and let $A_k(x) = \sum_{n=1}^{\infty} x^{\alpha_k(n)}/n^k$. Then $A_k(x)$ converges at a positive real number x if and only if $x < k^k - k + 1$.

Problem [49, 1988, A3]

Determine, with proof, the set of real numbers x for which

$$\sum_{n=1}^{\infty} \left(\frac{1}{n} \csc \frac{1}{n} - 1\right)^x$$

converges.

Answer: The series converges if and only if $x > 1/2$.

Solution. Define

$$a_n = \frac{1}{n} \csc \frac{1}{n} - 1 = \frac{1}{n \sin \frac{1}{n}} - 1.$$

Taking $t = 1/n$ in the inequality $0 < \sin t < t$ for $t \in (0, \pi)$, we obtain $a_n > 0$, so each term a_n^x of the series is defined for any real x. Using $\sin t = t - t^3/3! + O(t^5)$ as $t \to 0$, we have as $n \to \infty$,

$$a_n = \frac{1}{n\left(\frac{1}{n} - \frac{1}{6n^3} + O\left(\frac{1}{n^5}\right)\right)} - 1$$
$$= \frac{1}{1 - \frac{1}{6n^2} + O\left(\frac{1}{n^4}\right)} - 1$$
$$= \frac{1}{6n^2} + O\left(\frac{1}{n^4}\right).$$

In particular, if $b_n = 1/n^2$, then a_n^x / b_n^x has finite limit as $n \to \infty$, so by the Limit Comparison Test, $\sum_{n=1}^{\infty} a_n^x$ converges if and only if $\sum_{n=1}^{\infty} b_n^x = \sum_{n=1}^{\infty} n^{-2x}$ converges, which by the Integral Comparison Test holds if and only if $2x > 1$, i.e., $x > 1/2$.

Problem [49, 1988, B4]

Prove that if $\sum_{n=1}^{\infty} a_n$ is a convergent series of positive real numbers, then so is

$$\sum_{n=1}^{\infty} (a_n)^{n/(n+1)}.$$

Solution. If $a_n \geq 1/2^{n+1}$, then

$$a_n^{n/(n+1)} = \frac{a_n}{a_n^{1/(n+1)}} \leq 2a_n.$$

If $a_n \leq 1/2^{n+1}$, then $a_n^{n/(n+1)} \leq 1/2^n$. Hence

$$a_n^{n/(n+1)} \leq 2a_n + \frac{1}{2^n}.$$

Since $\sum_{n=1}^{\infty}(2a_n + 1/2^n)$ converges, $\sum_{n=1}^{\infty} a_n^{n/(n+1)}$ also converges by the Comparison Test.

Problem [55, 1994, A1]

Suppose that a sequence $a_1, a_2, a_3, \ldots$ satisfies $0 < a_n < a_{2n} + a_{2n+1}$ for all $n \geq 1$. Prove that series $\sum_{n=1}^{\infty} a_n$ diverges.

Solution. For $m \geq 1$, let $b_m = \sum_{i=2^{m-1}}^{2^m - 1} a_i$. Summing $a_n \leq a_{2n} + a_{2n+1}$ from $n = 2^{m-1}$ to $n = 2^m - 1$ yields $b_m \leq b_{m+1}$ for all $m \geq 1$. For any $t \geq 1$,

$$\sum_{n=1}^{2^t - 1} a_n = \sum_{m=1}^{t} b_m \geq t b_1 = t a_1,$$

which is unbounded as $t \to \infty$ since $a_1 > 0$, so $\sum_{n=1}^{\infty} a_n$ diverges.

Alternatively, assuming the series converges to a finite value L, we obtain the contradiction

$$\begin{aligned} L &= b_1 + (b_2 + b_3 + \cdots) \\ &\geq b_1 + (b_1 + b_2 + \cdots) \\ &= b_1 + L. \end{aligned}$$

Problem [60, 1999, A4]

Sum the series

$$\sum_{m=1}^{\infty} \sum_{n=1}^{\infty} \frac{m^2 n}{3^m (n 3^m + m 3^n)}.$$

Answer: The series converges to $9/32$.

Solution. Denote the series by S, and let $a_n = 3^n/n$. Note that

$$S = \sum_{m=1}^{\infty} \sum_{n=1}^{\infty} \frac{1}{a_m(a_m + a_n)} = \sum_{m=1}^{\infty} \sum_{n=1}^{\infty} \frac{1}{a_n(a_m + a_n)},$$

where the second equality follows by interchanging m and n. Thus

$$\begin{aligned} 2S &= \sum_m \sum_n \left(\frac{1}{a_m(a_m + a_n)} + \frac{1}{a_n(a_m + a_n)} \right) \\ &= \sum_m \sum_n \frac{1}{a_m a_n} = \left(\sum_{n=1}^{\infty} \frac{n}{3^n} \right)^2. \end{aligned}$$

Finally, if

$$A = \sum_{n=1}^{\infty} \frac{n}{3^n}$$

then

$$3A = \sum_{n=1}^{\infty} \frac{n}{3^{n-1}} = \sum_{n=0}^{\infty} \frac{n+1}{3^n}$$

so subtracting gives

$$2A = 1 + \sum_{n=1}^{\infty} \frac{1}{3^n} = \frac{3}{2}.$$

Hence

$$A = \frac{3}{4},$$

so $S = 9/32$.

Remark. This solution uses a trick that is powerful in general, namely interchanging the roles of the variables and combining the results to introduce a more symmetric expression, which is often more workable. It is important when doing such things to check that the series does in fact converge to something. Operating on a divergent series can sometimes give a convergent result. In this case, since the terms of the double series are positive, the partial sums are bounded above by $9/32$ (using the given argument), so that this series must be convergent. Since it is convergent, the analysis above applies, so it does sum to 9/32. In this case, there is no additional work to be done, but it is important to remember not to assume convergence *a priori*.

Remark (by Kedlaya, Poonen, Vakil) The value of A can also be derived by differentiating both sides of

$$\sum_{n=0}^{\infty} \frac{x^n}{3^n} = \frac{3}{3-x},$$

and then evaluating at $x = 1$.

Problem [61, 2000, A1]

Let A be a positive real number. What are the possible values of $\sum_{j=0}^{\infty} x_j^2$, given that $x_0, x_1, \ldots$ are positive numbers for which $\sum_{j=0}^{\infty} x_j = A$?

Answer: The possible values comprise the interval $(0, A^2)$.

Solution 1. Since all terms in the series are positive, we may rearrange terms to deduce

$$0 < \sum x_i^2 < \sum x_i^2 + \sum_{i<j} 2x_i x_j = \left(\sum x_i\right)^2 = A^2.$$

Thus it remains to show that each number in $(0, A^2)$ is a possible value of $\sum x_i^2$.

We use geometric series. Given $0 < r < 1$, there is a geometric series $\sum x_i$ with common ratio r and sum A: it has $x_0/(1-r) = A$ so $x_0 = (1-r)A$ and $x_j = r^j(1-r)A$. Then

$$\sum_{j=0}^{\infty} x_j^2 = \frac{x_0^2}{1-r^2} = \frac{1-r}{1+r} A^2.$$

To make this equal a given number $B \in (0, A^2)$, take $r = (A^2 - B)/(A^2 + B)$.

Solution 2. As in Solution 1, we prove $0 < \sum x_i^2 < A^2$, and it remains to show that each number in $(0, A^2)$ is a possible value of $\sum x_i^2$. There exists a series of positive numbers $\sum x_i$ with sum A. Then $x_0/2, x_0/2, x_1/2, x_1/2, x_2/2, \ldots$ also sums to A but its squares sum to half the previous sum of squares. Iterating shows that the sum of squares can be arbitrarily small.

Given any series $\sum x_i$ of positive terms with sum A, form a weighted average of it with the series $A + 0 + 0 + \cdots$; in other words, choose $t \in (0, 1)$ and define a new series of positive terms $\sum y_i$ by setting $y_0 = tx_0 + (1-t)A$ and $y_i = tx_i$ for $i \geq 1$. Then $\sum y_i = A$, and

$$\sum y_i^2 = t^2 \sum x_i^2 + 2t(1-t)x_0 A + (1-t)^2 A^2.$$

As t runs from 0 to 1, the Intermediate Value Theorem shows that this quadratic polynomial in t takes on all values strictly between $\sum x_i^2$ and A^2. Since $\sum x_i^2$ can be made arbitrarily small, any number $(0, A^2)$ can occur as $\sum y_i^2$.

Problem [62, 2001, B3]

For any positive integer n, let $\langle n \rangle$ denote the closest integer to $\sqrt{n}$. Evaluate

$$\sum_{n=1}^{\infty} \frac{2^{\langle n \rangle} + 2^{-\langle n \rangle}}{2^n}.$$

Answer. The series converges to 3.

Solution (by the authors). The trick is to regroup this sum based on the value of $\langle n \rangle$. First, we notice that

$$\begin{aligned}
\langle n \rangle = k &\Leftrightarrow k - 1/2 \leq \sqrt{n} < k + 1/2 \\
&\Leftrightarrow k^2 - k + 1/4 \leq n < k^2 + k + 1/4 \\
&\Leftrightarrow k^2 - k + 1 \leq n \leq k^2 + k.
\end{aligned}$$

Now we have

$$\begin{aligned}
\sum_{n=1}^{\infty} \frac{2^{\langle n \rangle} + 2^{-\langle n \rangle}}{2^n} &= \sum_{k=1}^{\infty} \sum_{n=k^2-k+1}^{k^2+k} \frac{2^k + 2^{-k}}{2^n} \\
&= \sum_{k=1}^{\infty} (2^k + 2^{-k}) \sum_{n=k^2-k+1}^{k^2+k} 2^{-n} \\
&= \sum_{k=1}^{\infty} (2^k + 2^{-k}) \left(2^{-k^2+k} - 2^{-k^2-k}\right) \\
&= \sum_{k=1}^{\infty} \left(2^{-k^2+2k} + 2^{-k^2} - 2^{-k^2} - 2^{-k^2-2k}\right) \\
&= \sum_{k=1}^{\infty} \left(2^{-k(k-2)} - 2^{-k(k+2)}\right) \\
&= \sum_{k=1}^{\infty} 2^{-k(k-2)} - \sum_{k=1}^{\infty} 2^{-k(k+2)}
\end{aligned}$$

$$= \sum_{k=1}^{\infty} 2^{-k(k-2)} - \sum_{k=3}^{\infty} 2^{-k(k-2)}$$

$$= \sum_{k=1}^{2} 2^{-k(k-2)} = 2 + 1 = 3.$$

Note that

$$\sum_{n=1}^{\infty} \frac{2^{\langle n \rangle} + 2^{-\langle n \rangle}}{2^n} = \sum_{n=1}^{\infty} \frac{1}{2^{n-\langle n \rangle}} + \sum_{n=1}^{\infty} \frac{1}{2^{n+\langle n \rangle}}.$$

Problem [63, 2002, A6]

Fix an integer $b \geq 2$. Let $f(1) = 1$, $f(2) = 2$, and for each $n \geq 3$, define $f(n) = nf(d)$, where d is the number of base-b digits of n. For which values of b does

$$\sum_{n=1}^{\infty} \frac{1}{f(n)}$$

converge?

Answer: The series converges only for $b = 2$.

Solution (by the authors). As in the previous problem, the trick is a clever regrouping. Let $b > 2$, and suppose for the sake of contradiction that the series converges. Notice that $f(n) = nf(d)$ for $b^{d-1} \leq n \leq b^d - 1$. Then we have, comparing to a Riemann sum at the appropriate moment,

$$\sum_{n=1}^{\infty} \frac{1}{f(n)} = \sum_{d=1}^{\infty} \sum_{n=b^{d-1}}^{b^d-1} \frac{1}{nf(d)}$$

$$= \sum_{d=1}^{\infty} \frac{1}{f(d)} \sum_{n=b^{d-1}}^{b^d-1} \frac{1}{n}$$

$$\geq \sum_{d=1}^{\infty} \frac{1}{f(d)} \int_{b^{d-1}}^{b^d} \frac{dx}{x}$$

$$= \sum_{d=1}^{\infty} \frac{1}{f(d)} [\log(b^d) - \log(b^{d-1})]$$

$$= \log b \sum_{d=1}^{\infty} \frac{1}{f(d)}$$

$$> \sum_{d=1}^{\infty} \frac{1}{f(d)},$$

since $\log b \geq \log 3 > 1$. But this is a contradiction, since we supposed the series to converge to a positive real number.

Now take $b = 2$. First, by a similar line of reasoning to the above, we have for each $k \geq 4$,

$$\begin{aligned}
\sum_{n=4}^{2^k-1} \frac{1}{f(n)} &= \sum_{d=3}^{k} \sum_{n=2^{d-1}}^{2^d-1} \frac{1}{nf(d)} \\
&= \sum_{d=3}^{k} \frac{1}{f(d)} \sum_{n=2^{d-1}}^{2^d-1} \frac{1}{n} \\
&\leq \sum_{d=3}^{k} \left(\frac{1}{f(d)} \int_{2^{d-1}}^{2^d} \frac{dx}{x} + \frac{1}{2^d} \right) \\
&= \sum_{d=3}^{k} \left(\frac{1}{f(d)} [\log(2^d) - \log(2^{d-1})] + \frac{1}{2^d} \right) \\
&= \sum_{d=3}^{k} \left(\frac{\log 2}{f(d)} + \frac{1}{2^d} \right) \\
&\leq \log 2 \sum_{d=4}^{k} \frac{1}{f(d)} + \log 2 \frac{1}{f(3)} + \frac{1}{4} \\
&= \log 2 \sum_{n=4}^{k} \frac{1}{f(d)} + \frac{\log 2}{6} + \frac{1}{4}.
\end{aligned}$$

Now, write

$$S(k) = \sum_{n=4}^{k} \frac{1}{f(n)}.$$

We have just proven that

$$S(2^k - 1) \leq \log 2(S(k) + 1/6) + 1/4.$$

We are then led to consider the equation

$$S = \log 2(S + 1/6) + (1/4)$$

which has solution $S = 1/(12 - 12 \log 2) = 0.4294485 \ldots$. Notice that $S(4) = 1/f(4) = 1/8 < 0.4294485 \ldots$. By a straightforward induction we can show that whenever $S(k) < 0.4294485 \ldots$, so also $S(2^k - 1) < 0.4294485$. Thus $S(k) < 0.4294485 \ldots$ for arbitrarily large k, so that in fact, since the $S(k)$ are certainly increasing (the terms of our series are positive), $S(k) < 0.4294485 \ldots$. By adding back the first three terms of our series, we see that our original series has bounded partial sums, so it converges.

6

Final Diversions

We hope that you have found the earlier chapters of this book to be enlightening and enjoyable. As a parting gift to you, we provide this concluding chapter as a sort of dessert, which we believe you will find entertaining and worthwhile. Here we include a few puzzles with solutions. Each puzzle has a solution that in some way involves reasoning about infinite series. A number of pictorial proofs of basic facts about infinite series are included as well. The results will be familiar to you by this point and the pictures may be illuminating. Additionally, we provide a collection of fallacious "proofs" collected from math club presentations and from memorable discussions with colleagues while enjoying a cup of coffee. References labeled FFF refer to Fallacies, Flaws and Flimflam as offered in the *College Mathematics Journal* (CMJ) and the MAA book by Edward Barbeau entitled *Mathematical Fallacies, Flaws and Flimflam.* We invite the reader to determine what is wrong with each of these arguments.

6.1 Puzzles

Puzzle 1

(Note: This puzzle is often referred to as the "pebbling problem.") Consider an infinite grid of squares with one corner (i.e., a single quadrant). Six of the squares are colored, as indicated in the figure. At the beginning, these six colored squares are each occupied by a

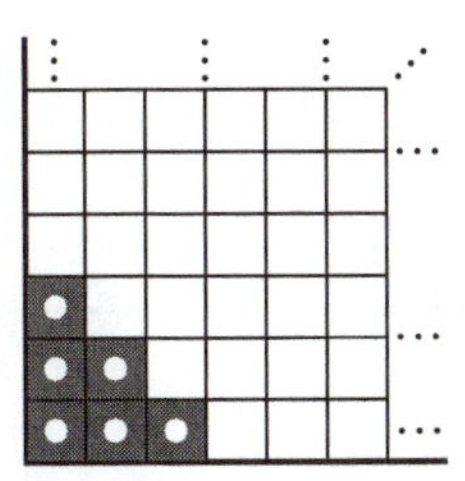

FIGURE 6.1
Setup for the Pebbling Problem

single stone (or pebble). Now, we are allowed to remove any stone, provided we replace it by two new stones in the square directly above and in the square directly to the right. If either of these squares are already occupied, the stone cannot be removed at that time.

1. Is it possible, by these operations, to evacuate all six of the colored squares?
2. Suppose instead that only the corner square is occupied by a stone. In this case, is it possible to evacuate the six colored squares?

Puzzle 2

Now consider an infinite checkerboard, this time infinite in all directions. On this board, there is drawn a horizontal line dividing the checkerboard into an upper and lower half. The following solitaire game is played. The player begins by placing any finite number of checkers, at most one to a square, in the lower half of the board. Then the game is played like peg solitaire. Any checker may jump over an adjacent (but not diagonally) checker and land in the square just beyond, provided the destination square is unoccupied. The jumped-over checker is removed. The goal of the game is to get a checker as far into the upper half of the board as possible given the starting configuration. For each positive integer n, let c_n be the minimum number of starting checkers necessary to place a checker in the nth row above the line. For example, $c_1 = 2$, $c_2 = 4$, as indicated in Figures 6.2 and 6.3. How does the sequence c_n increase as n gets large?

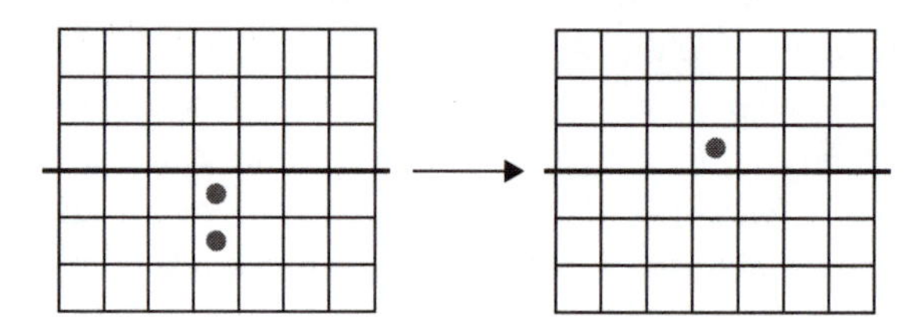

FIGURE 6.2
Solution for $c_1 = 2$

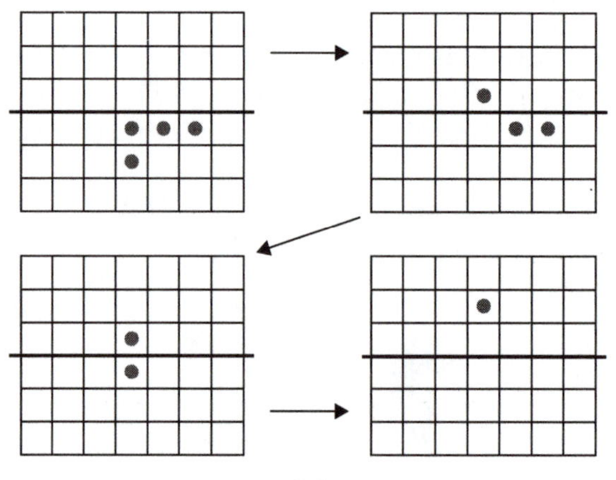

FIGURE 6.3
Solution for $c_2 = 4$

Puzzle 3

Desert Dan is trying to cross a 500-mile desert in a jeep. Unfortunately, his jeep can hold only enough fuel to travel 50 miles. However, Desert Dan has an infinite collection of containers that he can use to leave caches of fuel for himself to use on a later trip. Further, on the entering edge of the desert there is a boundless supply of fuel, so that Desert Dan can return to the entering edge of the desert as often as he wishes. Can he cross the desert?

Solutions for these puzzles appear at the end of this chapter.

6.2 Visuals

Proofs without Words (PWWs) are what Martin Gardner describes as "look-see" diagrams that assist in illustrating the beauty and truth of mathematical results or theorems. What follows is a set of diagrams (PWWs) as offered in the Mathematical Association of America publications entitled *The American Mathematical Monthly* (AMM), *Mathematics Magazine* (MM), and *The College Mathematics Journal* (CMJ). Nearly all of these diagrams also appear in the two MAA published books by Roger B. Nelsen entitled *Proofs without Words* and *Proofs without Words II*. We also make available three additional visuals, not PWWs, but nevertheless very similar in spirit. We list here the title for each visual and credit the individual(s) who made the result available. The source for the original publication is also indicated.

Can You Sum This Familiar Series?

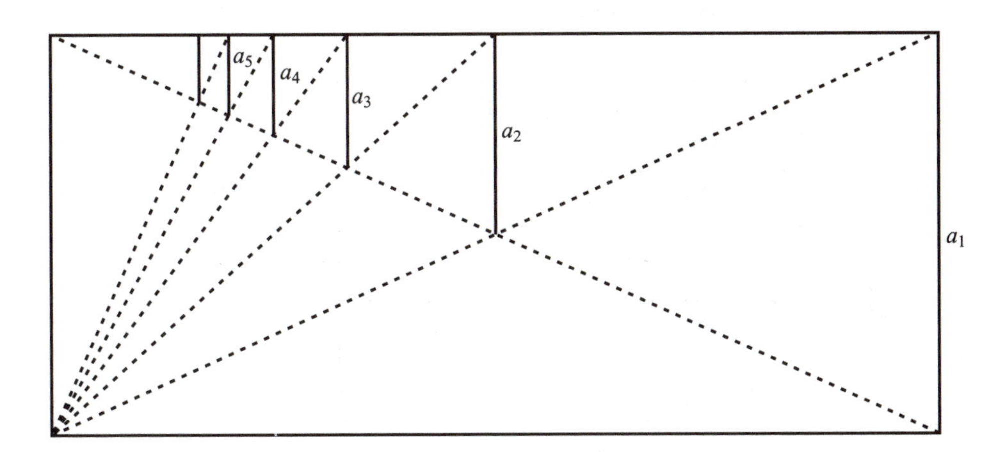

— Dennis Gittinger
Reprinted from *Coll. Math. J.*, 28, (Nov., 1997), 393

Geometric Sums

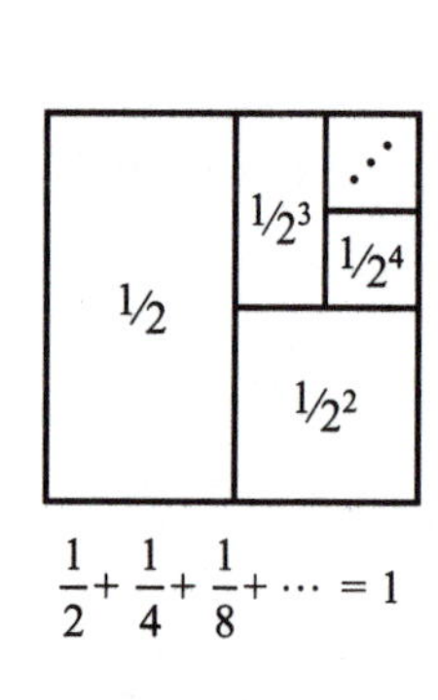

$$\frac{1}{2}+\frac{1}{4}+\frac{1}{8}+\cdots=1$$

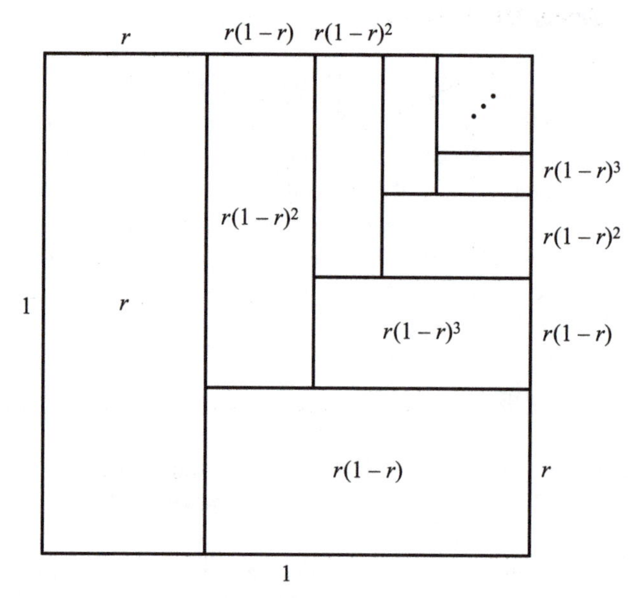

$$r+r(1-r)+r(1-r)^2+\cdots=1$$

— Warren Page
Reprinted from *Math. Mag.*, 54, (Sept., 1981), 201

Geometric Series I

$$\sum_{n=0}^{\infty} ar^n = \frac{a}{1-r}$$

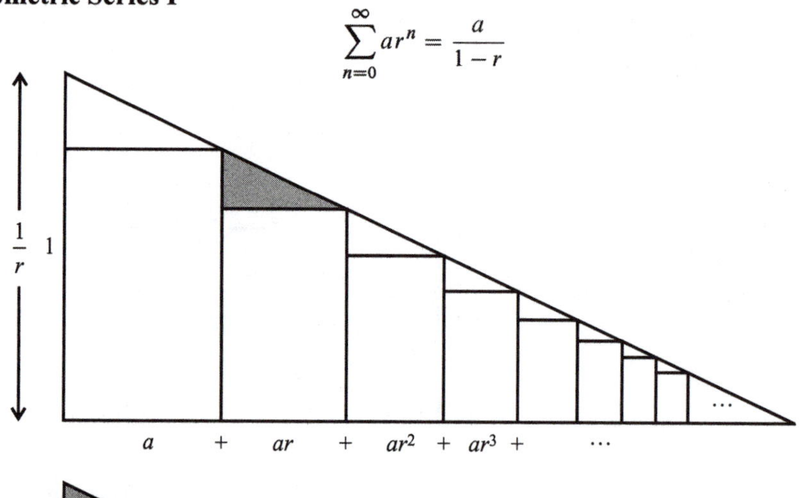

$$\frac{a+ar+ar^2+ar^3+\cdots}{1/r}=\frac{ar}{1-r}$$

— J. H. Webb
Reprinted from *Math.Mag.*, 60, (June, 1987), 177

Geometric Series II

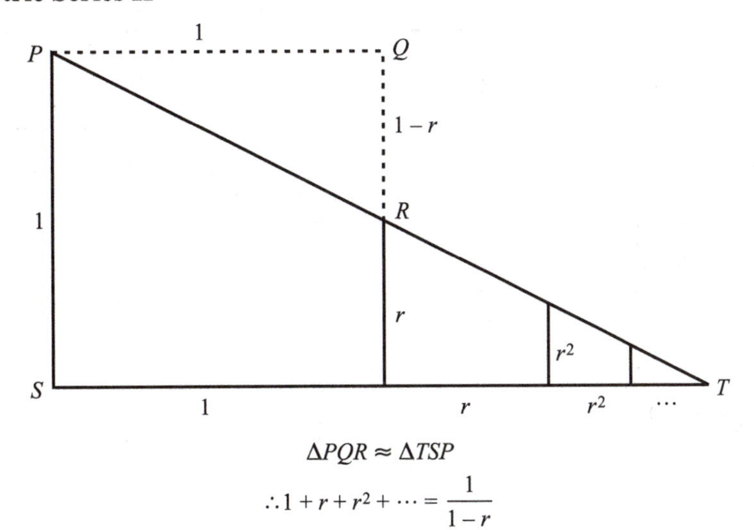

$$\Delta PQR \approx \Delta TSP$$

$$\therefore 1 + r + r^2 + \cdots = \frac{1}{1-r}$$

— Benjamin G. Klein and Irl C. Bivens
Reprinted from *Math. Mag.*, 61, (Oct. 1988), 219

Geometric Series III

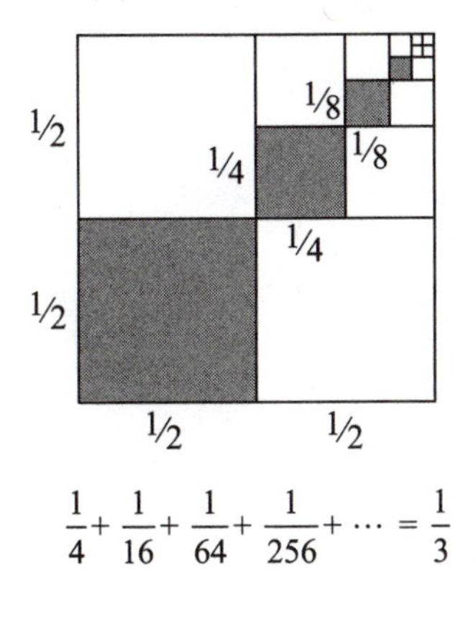

$$\frac{1}{4} + \frac{1}{16} + \frac{1}{64} + \frac{1}{256} + \cdots = \frac{1}{3}$$

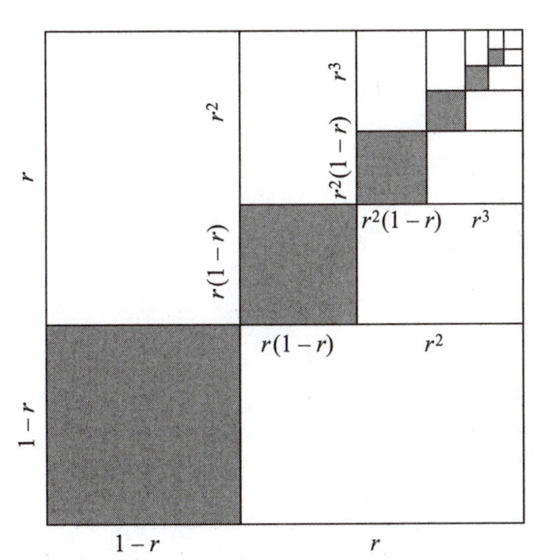

$$(1-r)^2 + r^2(1-r)^2 + r^4(1-r)^2 + \cdots = \frac{(1-r)^2}{(1-r)^2 + 2r(1-r)} = \frac{1-r}{1+r}$$

$$1 + r^2 + r^4 + \cdots = \frac{1}{1-r^2}$$

$$a + ar + ar^2 + \cdots = \frac{a}{1-r}$$

— Sunday A. Ajose
Reprinted from *Math. Mag.*, 67, (June 1994), 230

Geometric Series IV

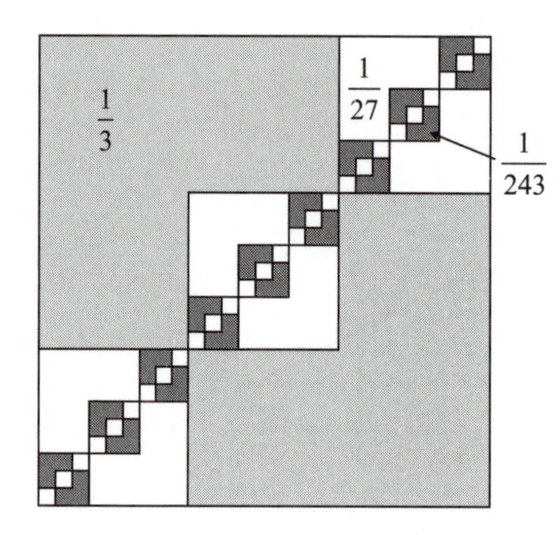

$$2\left(\frac{1}{3}+3\cdot\frac{1}{27}+9\cdot\frac{1}{243}+\cdots\right)=1$$

$$2\sum_{n=1}^{\infty}\frac{1}{3^n}=1$$

$$\sum_{n=1}^{\infty}\frac{1}{3^n}=\frac{1}{2}$$

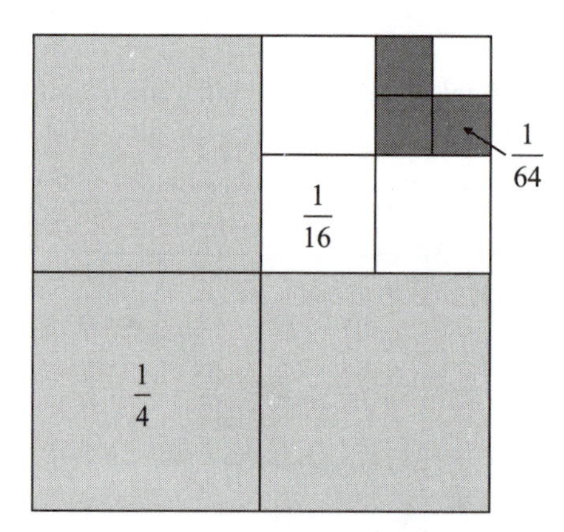

$$3\sum_{n=1}^{\infty}\frac{1}{4^n}=1$$

$$\sum_{n=1}^{\infty}\frac{1}{4^n}=\frac{1}{3}$$

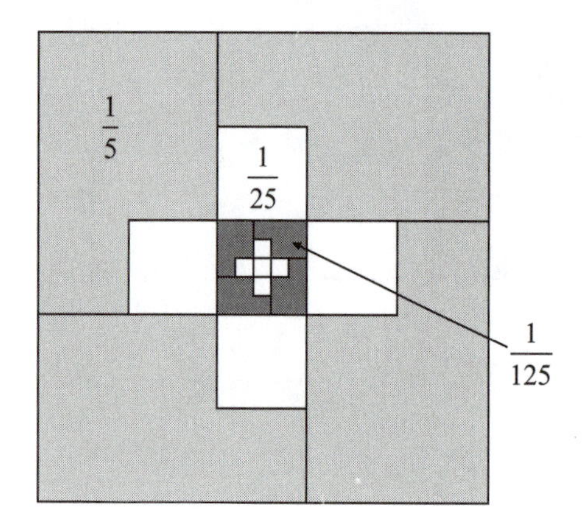

$$4\sum_{n=1}^{\infty}\frac{1}{5^n}=1$$

$$\sum_{n=1}^{\infty}\frac{1}{5^n}=\frac{1}{4}$$

— Elizabeth M. Markham

Reprinted from *Math. Mag.*, 66, (Oct., 1993), 242

Gabriel's Staircase

$$\sum_{k=1}^{\infty} k r^k = \frac{r}{(1-r)^2} \quad \text{for } 0 < r < 1$$

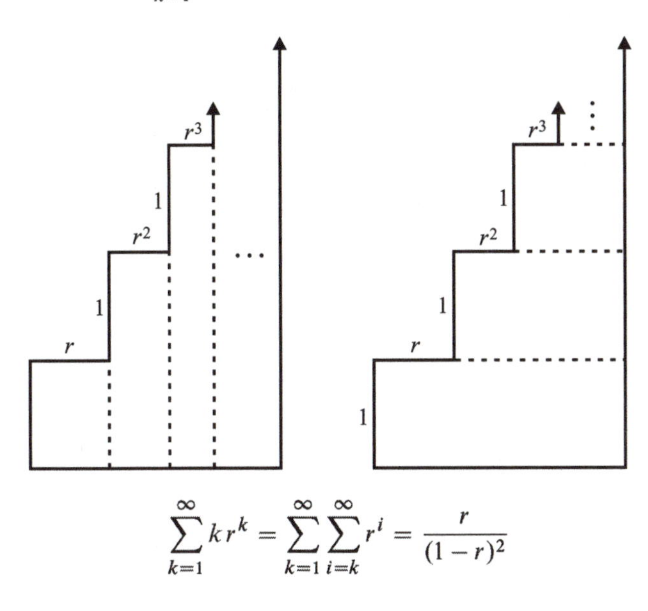

$$\sum_{k=1}^{\infty} k r^k = \sum_{k=1}^{\infty} \sum_{i=k}^{\infty} r^i = \frac{r}{(1-r)^2}$$

— Stuart G. Swain
Reprinted from *Math. Mag.*, 67, (June, 1994), 209

A Geometric Series

$$\frac{1}{4} + \left(\frac{1}{4}\right)^2 + \left(\frac{1}{4}\right)^3 + \cdots = \frac{1}{3}$$

— Rick Mabry
Reprinted from *Math. Mag.*, 72, (Feb., 1999), 63

Differentiated Geometric Series

$$1 + 2\left(\frac{1}{2}\right) + 3\left(\frac{1}{4}\right) + 4\left(\frac{1}{8}\right) + \cdots = 4$$

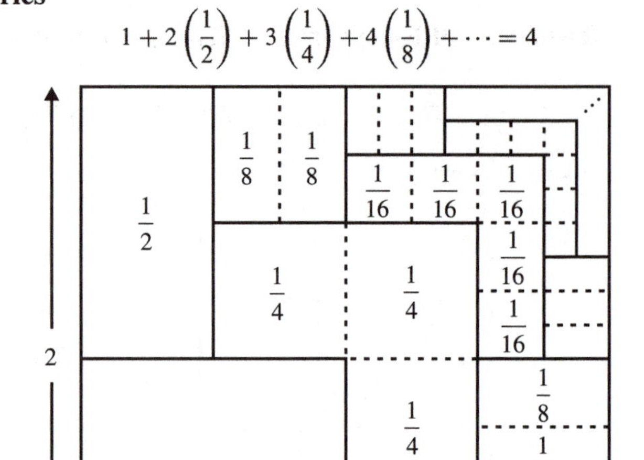

$$1 + 2r + 3r^2 + 4r^3 + \cdots = \left(\frac{1}{1-r}\right)^2, \quad 0 \le r < 1$$

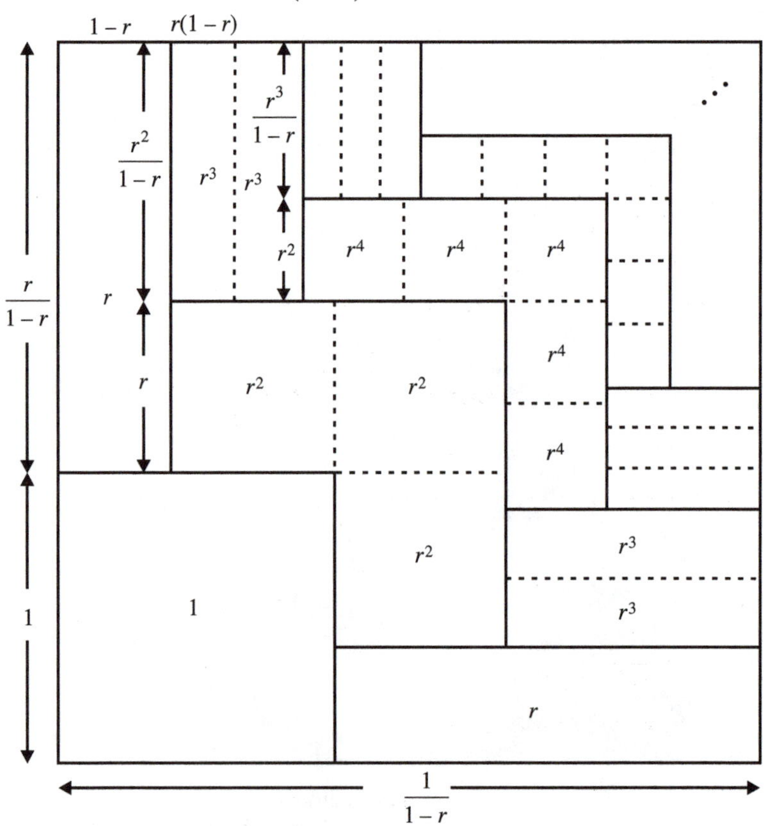

— Roger B. Nelsen

Reprinted from *Math. Mag.*, 62, (Dec., 1989), 332–333

The Series of Reciprocals of Triangular Numbers

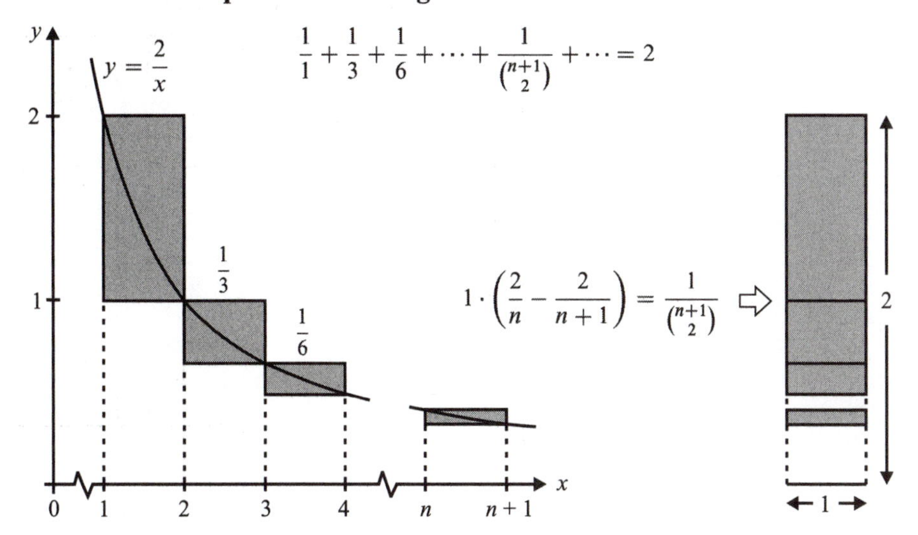

— Roger B. Nelsen
Reprinted from *Math. Mag.*, 64, (June 1991), 167

The Alternating Harmonic Series

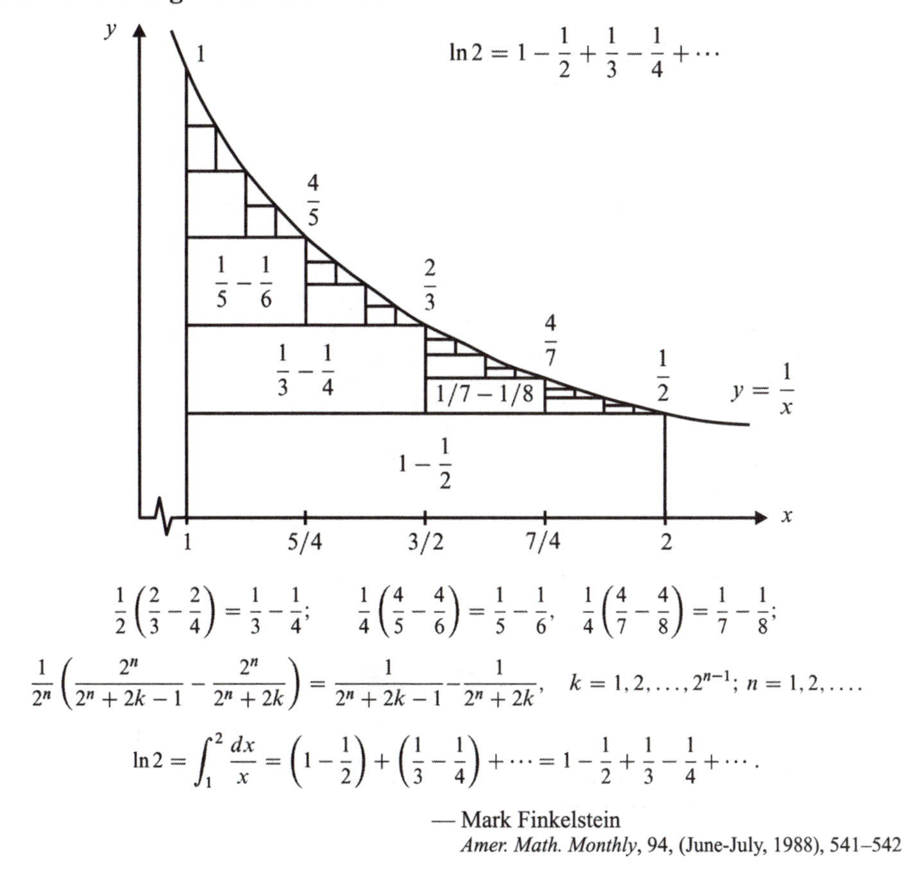

$$\frac{1}{2}\left(\frac{2}{3}-\frac{2}{4}\right)=\frac{1}{3}-\frac{1}{4};\qquad \frac{1}{4}\left(\frac{4}{5}-\frac{4}{6}\right)=\frac{1}{5}-\frac{1}{6},\quad \frac{1}{4}\left(\frac{4}{7}-\frac{4}{8}\right)=\frac{1}{7}-\frac{1}{8};$$

$$\frac{1}{2^n}\left(\frac{2^n}{2^n+2k-1}-\frac{2^n}{2^n+2k}\right)=\frac{1}{2^n+2k-1}-\frac{1}{2^n+2k},\quad k=1,2,\ldots,2^{n-1};\ n=1,2,\ldots.$$

$$\ln 2=\int_1^2\frac{dx}{x}=\left(1-\frac{1}{2}\right)+\left(\frac{1}{3}-\frac{1}{4}\right)+\cdots=1-\frac{1}{2}+\frac{1}{3}-\frac{1}{4}+\cdots.$$

— Mark Finkelstein
Amer. Math. Monthly, 94, (June-July, 1988), 541–542

An Alternating Series

$$\frac{1}{2}-\frac{1}{4}+\frac{1}{8}-\frac{1}{16}+\frac{1}{32}-\frac{1}{64}+\cdots=\frac{1}{3}$$

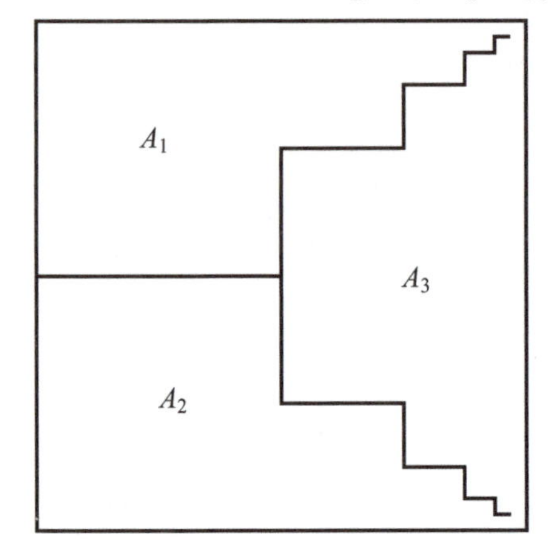

$$\begin{aligned} A_1 &= \frac{1}{2}-\frac{1}{4}+\frac{1}{8}-\frac{1}{16}+\frac{1}{32}-\frac{1}{64}+\cdots, \\ A_1 &= A_2 = A_3, \\ A_1 &+ A_2 + A_3 = 1, \\ \therefore A_1 &= \frac{1}{3} \end{aligned}$$

— James O. Chilaka
Reprinted from *Math. Mag.*, 69, (Dec., 1996), 355–356

A Generalized Geometric Series

Let $\{k_1, k_2, k_3, \ldots\}$ be a sequence of integers, each of which is at least 2. Then

$$\frac{k_1-1}{k_1}+\frac{k_2-1}{k_2k_1}+\frac{k_3-1}{k_3k_2k_1}+\cdots=1.$$

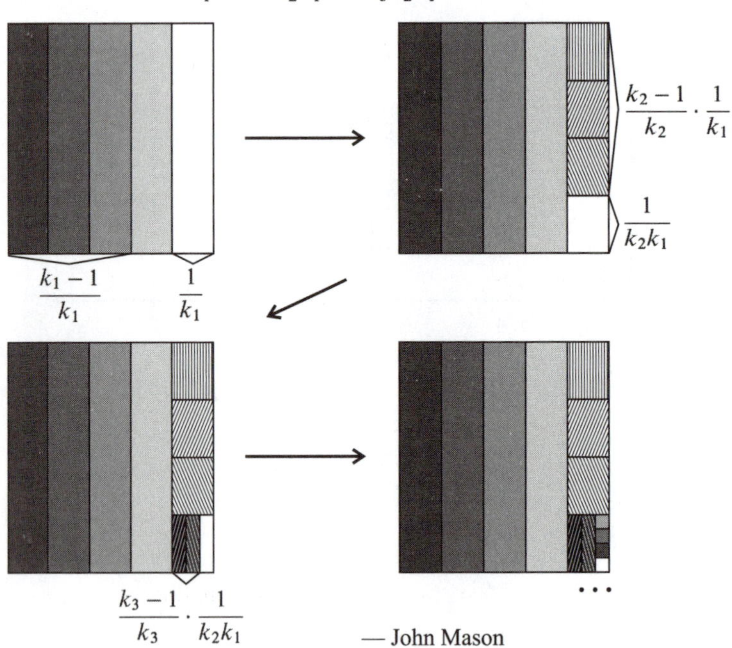

— John Mason
Reprinted from *Coll. Math. J.*, 26, (Nov., 1995), 381

Divergence of a Series

$$n > 1 \Rightarrow \sum_{k=1}^{n} \frac{1}{\sqrt{k}} > \sqrt{n}$$

$k > 1 \Rightarrow$

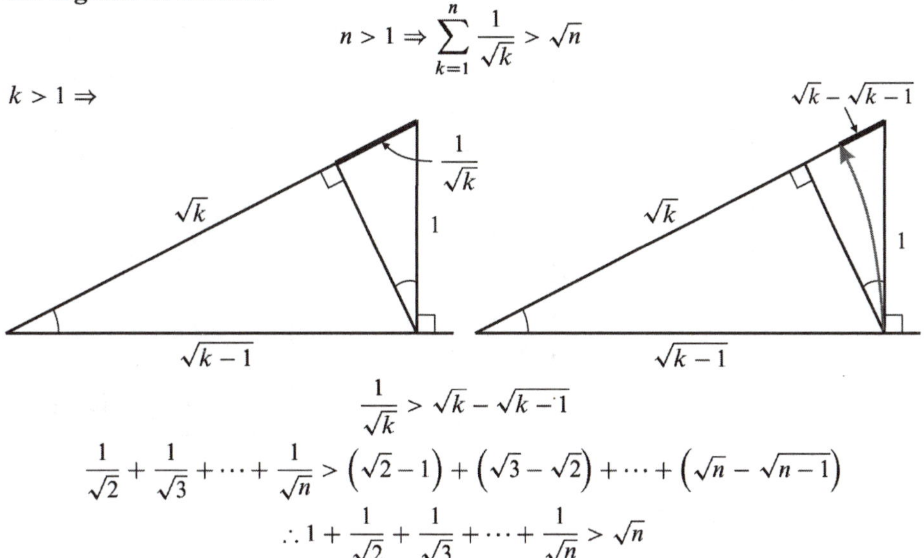

$$\frac{1}{\sqrt{k}} > \sqrt{k} - \sqrt{k-1}$$

$$\frac{1}{\sqrt{2}} + \frac{1}{\sqrt{3}} + \cdots + \frac{1}{\sqrt{n}} > \left(\sqrt{2}-1\right) + \left(\sqrt{3}-\sqrt{2}\right) + \cdots + \left(\sqrt{n}-\sqrt{n-1}\right)$$

$$\therefore 1 + \frac{1}{\sqrt{2}} + \frac{1}{\sqrt{3}} + \cdots + \frac{1}{\sqrt{n}} > \sqrt{n}$$

Sydney H. Kung
Reprinted from *Coll. Math. J.*, 26, (Sept., 1995), 301

Sum the Alternating Harmonic Series

$$\lim_{n\to\infty} \sum_{k=1}^{2n} (-1)^{k+1} \frac{1}{k} = \ln 2.$$

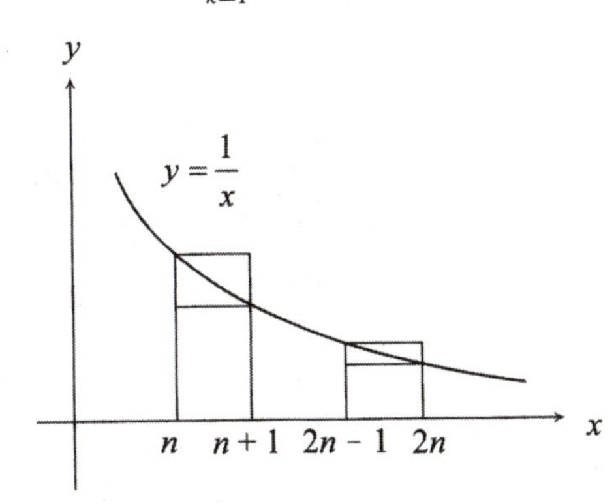

$$\sum_{k=1}^{2n} (-1)^{k+1} \frac{1}{k} = \sum_{k=1}^{2n} \frac{1}{k} - 2\sum_{k=1}^{n} \frac{1}{2k} = \sum_{k=n+1}^{2n} \frac{1}{k} < \int_{n}^{2n} \frac{dx}{x} < \sum_{k=n}^{2n-1} \frac{1}{k}$$

$$= \sum_{k=1}^{2n-1} \frac{1}{k} - 2\sum_{k=1}^{n-1} \frac{1}{2k} = \sum_{n=1}^{2n-1} (-1)^{k+1} \frac{1}{k}.$$

Norman Schaumberger
Reprinted from *Coll. Math. J.*, 33, (Jan., 2002), 23

An Arctangent Identity and Series

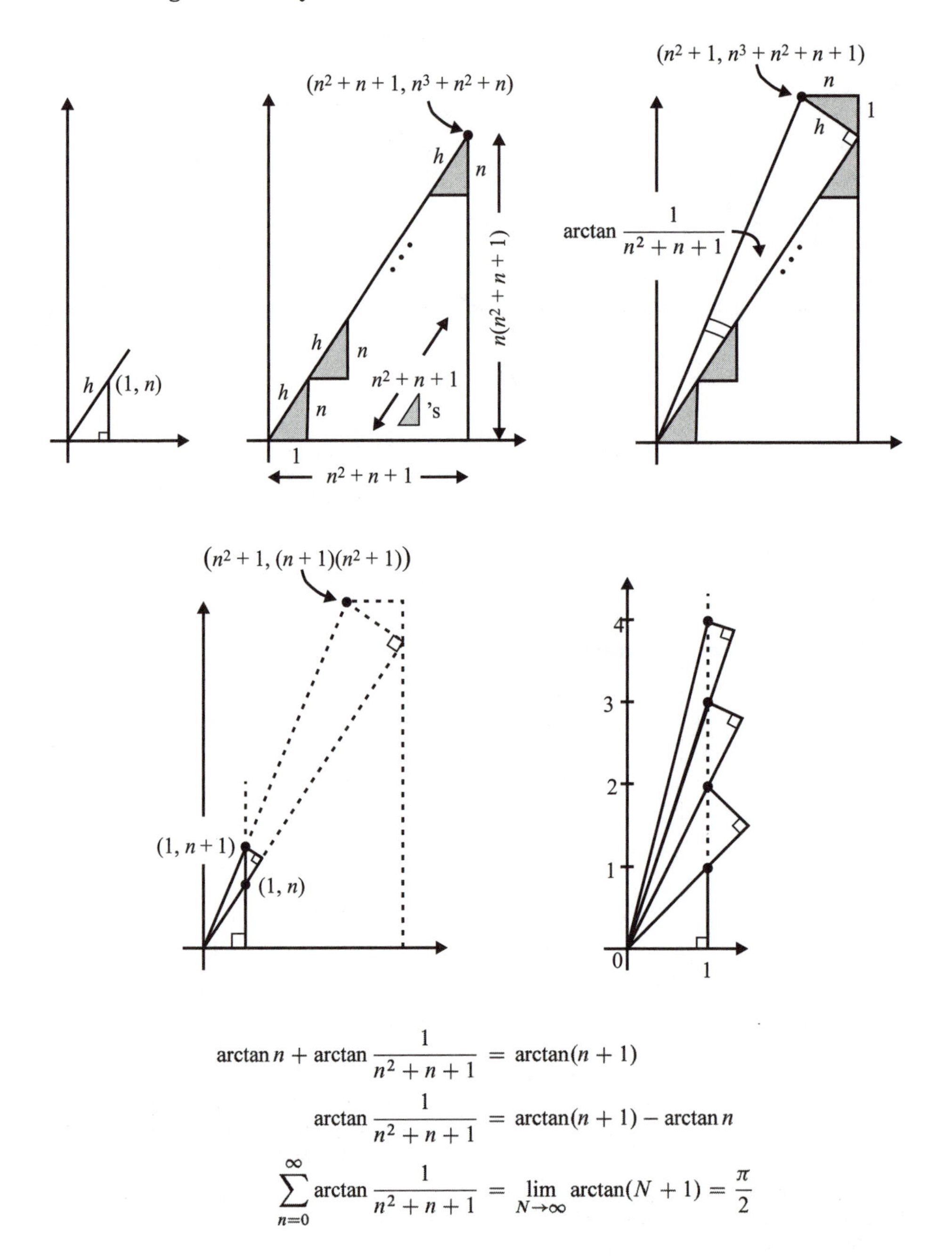

$$\arctan n + \arctan \frac{1}{n^2+n+1} = \arctan(n+1)$$

$$\arctan \frac{1}{n^2+n+1} = \arctan(n+1) - \arctan n$$

$$\sum_{n=0}^{\infty} \arctan \frac{1}{n^2+n+1} = \lim_{N\to\infty} \arctan(N+1) = \frac{\pi}{2}$$

— Roger B. Nelsen

Reprinted from Roger Nelsen, *Proofs without Words*, MAA, 1993. Originally appeared in *Math. Mag.*, 64, (Oct. 1991), 241

Geometric Series

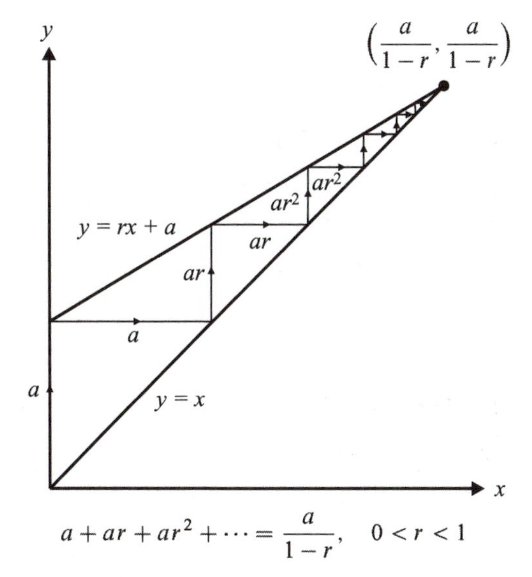

$$a + ar + ar^2 + \cdots = \frac{a}{1-r}, \quad 0 < r < 1$$

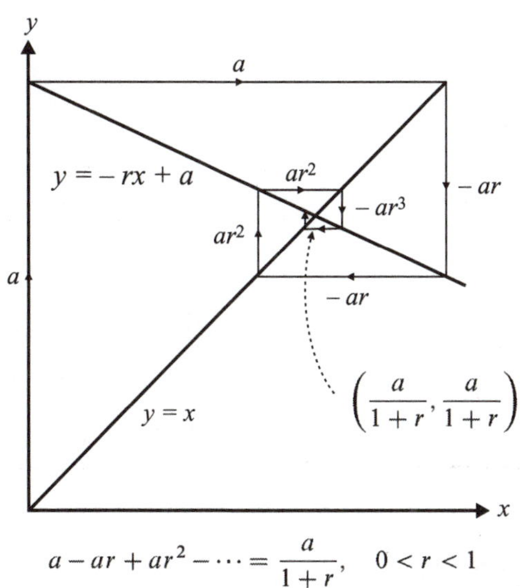

$$a - ar + ar^2 - \cdots = \frac{a}{1+r}, \quad 0 < r < 1$$

— The VIEWPOINTS 2000 Group*

Reprinted from *Math. Mag.*, 74, (Oct., 2001), 320

* The VIEWPOINTS 2000 Group is a subset of the participants in the NSF-sponsored VIEWPOINTS Mathematics and Art workshop, held at Franklin & Marshall College in June, 2000:

Marion Cohen, Drexel University, Philadelphia, PA 19104
Douglas Ensley, Shippensburg University, Shippensburg, PA 17257
Marc Frantz, Indiana University, Bloomington, IN 47405
Patricia Hauss, Arapahoe Community College, Littleton, CO 80160
Judy Kennedy, University of Delaware, Newark, DE 19716
Kerry Mitchell, University of Advancing Computer Technology, Tempe, AZ 85283
Patricia Oakley, Goshen College, Goshen, IN 46526

The Sum Is One

Many calculus texts contain the following telescoping series,

$$\sum_{n=1}^{\infty} \frac{1}{n(n+1)} = \sum_{n=1}^{\infty} \left(\frac{1}{n} - \frac{1}{n+1} \right) = 1. \tag{1}$$

An interesting geometric realization of this result can be obtained from the fact that the family of curves $\{x^n\}$ partitions the unit square into an infinite number of regions with total area 1. First, find the area between the curves $y = x^{n-1}$ and $y = x^n$ over the interval [0, 1]:

$$\int_0^1 (x^{n-1} - x^n)dx = \int_0^1 x^{n-1}\,dx - \int_0^1 x^n\,dx = \frac{1}{n} - \frac{1}{n+1} = \frac{1}{n(n+1)}. \tag{2}$$

Summing each side of (2) produces the result (1).

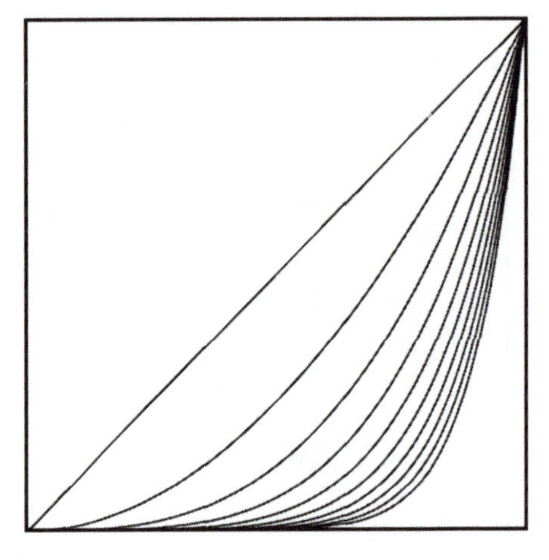

—John H. Mathews
Reprinted from *Coll. Math. J.* 22, (Sept., 1991), 322

6.3 Fallacious Proofs

There is an old saying to the effect that we can often learn more from our mistakes than from our successes. It is in this spirit that we offer this section for your enjoyment and your enlightment.

Fallacy 1

Let $u = 1 + \frac{1}{3} + \frac{1}{5} + \frac{1}{7} + \cdots$, and $v = \frac{1}{2} + \frac{1}{4} + \frac{1}{6} + \frac{1}{8} + \cdots$. Multiplying term by term,

$$2v = \frac{1}{1} + \frac{1}{2} + \frac{1}{3} + \frac{1}{4} + \frac{1}{5} + \frac{1}{6} + \cdots = u + v.$$

That is, $2v = u + v$, so that $u = v$, and hence $u - v = 0$.

On the other hand,

$$u - v = \left(1 - \frac{1}{2}\right) + \left(\frac{1}{3} - \frac{1}{4}\right) + \left(\frac{1}{5} - \frac{1}{6}\right) + \cdots$$

is a sum of positive terms, so it is also positive. Thus $0 = u - v > 0$, hence zero is positive.

Fallacy 2

Let $S = 1 + 2 + 4 + 8 + 16 + \cdots$. Clearly $S > 0$. Multiplying by 2, $2S = 2 + 4 + 8 + 16 + 32 + \cdots = S - 1$. So $2S = S - 1$, and $S = -1$. Then $-1 > 0$.

Fallacy 3

By long division, we arrive at the following power series:

$$\frac{1}{1+x} = 1 - x + x^2 - x^3 + x^4 - x^5 + \cdots$$

$$\frac{1}{1+x+x^2} = 1 - x + x^3 - x^4 + x^6 - x^7 + \cdots$$

$$\frac{1}{1+x+x^2+x^3} = 1 - x + x^4 - x^5 + x^8 - x^9 + \cdots$$

$$\frac{1}{1+x+x^2+x^3+x^4} = 1 - x + x^5 - x^6 + x^{10} - x^{11} + \cdots$$

$$\vdots$$

Now set $x = 1$. Then the right side of every equation is now the same, namely $1 - 1 + 1 - 1 + 1 - 1 + \cdots$. Thus the left sides must also be the same, that is $1/2 = 1/3 = 1/4 = 1/5 = 1/6 = \cdots = 1/n$ for all n.

Fallacy 4

Let $S = 2 - 2 + 2 - 2 + 2 - 2 + \cdots$. Then we have all of the following.

$$\begin{aligned}
S &= (2-2) + (2-2) + (2-2) + (2-2) + \cdots \\
&= 0 + 0 + 0 + 0 + \cdots = 0. \\
S &= 2 - (2-2) - (2-2) - (2-2) - (2-2) - \cdots \\
&= 2 - 0 - 0 - 0 - 0 - \cdots = 2. \\
S &= 2 - (2-2) - (2-2) - (2-2) - (2-2) - \cdots \\
&= 2 - S \\
&\Rightarrow 2S = 2 \Rightarrow S = 1.
\end{aligned}$$

That is, $0 = 1 = 2$.

Fallacy 5

Let $S = 1 - 2 + 4 - 8 + 16 - 32 + \cdots$. Then we have all of the following.

$$\begin{aligned} S &= 1 - 2 + 4 - 8 + 16 - 32 + \cdots \\ &= 1 - 2(1 - 2 + 4 - 8 + 16 - 32 + \cdots) = 1 - 2S \\ &\Rightarrow 3S = 1 \Rightarrow S = 1/3. \\ S &= 1 + (-2 + 4) + (-8 + 16) + (-32 + 64) + \cdots \\ &= 1 + 2 + 8 + 32 + \cdots = \infty. \\ S &= (1 - 2) + (4 - 8) + (16 - 32) + \cdots \\ &= -1 - 4 - 16 - \cdots = -\infty. \end{aligned}$$

That is, $1/3 = \infty = -\infty$.

Fallacy 6

It can be easily shown that

$$\sum_{n=1}^{\infty} \frac{1}{(2n-1)(2n+1)} = \frac{1}{3} + \frac{1}{15} + \frac{1}{35} + \frac{1}{63} + \cdots$$

converges. Call its sum S.

On the one hand,

$$S = \left(\frac{1}{1} - \frac{2}{3}\right) + \left(\frac{2}{3} - \frac{3}{5}\right) + \left(\frac{3}{5} - \frac{4}{7}\right) + \cdots.$$

By telescoping, $S = 1$.

On the other hand,

$$S = \left(\frac{1}{2} - \frac{1}{6}\right) + \left(\frac{1}{6} - \frac{1}{10}\right) + \left(\frac{1}{10} - \frac{1}{14}\right) + \cdots.$$

By telescoping, $S = 1/2$.

That is, 1/2=1.

Fallacy 7

This is the ultimate fallacy. Here we "prove" that any series whatsoever, convergent or divergent, in "fact" converges to any preselected number.

Let $\sum_{n=1}^{\infty} a_n$ be any series, and let S be any number. Then we have the following by simple rules of algebra.

$$\begin{aligned}
a_1 &= S - (-a_1 + S) \\
a_2 &= (-a_1 + S) - (-a_1 - a_2 + S) \\
a_3 &= (-a_1 - a_2 + S) - (-a_1 - a_2 - a_3 + S) \\
a_4 &= (-a_1 - a_2 - a_3 + S) - (-a_1 - a_2 - a_3 - a_4 + S) \\
&\vdots
\end{aligned}$$

When we sum these equations, the left sides give us back the original series. The right sides, however, telescope, with every term appearing both positively and negatively except for the very first. Thus, as claimed,

$$\sum_{n=1}^{\infty} a_n = S.$$

6.4 Fallacies, Flaws and Flimflam

FFF #20 A Power Series Representation [CMJ, 21:3, p. 217]

Problem. Expand about the origin $f(x) = (1+x^2)/(1-x^2)$. Note that $f(0) = 1$.

Solution. By the quotient rule, we find that

$$f'(x) = 2\left[\frac{2x}{(1-x^2)^2}\right] = 2\left[\frac{1}{1-x^2}\right]',$$

whence $f(x) = 2(1-x^2)^{-1} = 2(1+x^2+x^4+x^6+\cdots)$ (so that, in particular, $f(0) = 2$).

FFF #28 More Fun with Series [CMJ, 21:5, pp. 395–396]

$$\begin{aligned}
\frac{1}{1+t} &= \frac{1}{1-t^2} - t\frac{1}{1-t^4} - t^3\frac{1}{1-t^4} \\
&= 1 - t - t^3 + t^2 - t^5 - t^7 + t^4 - t^9 - t^{11} + \cdots
\end{aligned}$$

for $|t| < 1$. Integrating both sides between 0 and x yields

$$\log(1+x) = x - \frac{x^2}{2} - \frac{x^4}{4} + \frac{x^3}{3} - \frac{x^6}{6} - \frac{x^8}{8} + \cdots \quad (|x| < 1).$$

Taking the limit as x tends to 1 and invoking Abel's theorem, we obtain

$$\log 2 = 1 - \frac{1}{2} - \frac{1}{4} + \frac{1}{3} - \frac{1}{6} - \frac{1}{8} + \cdots.$$

However, grouping terms in the series yields

$$\left(1-\frac{1}{2}\right)-\frac{1}{4}+\left(\frac{1}{3}-\frac{1}{6}\right)-\frac{1}{8}+\cdots=\frac{1}{2}-\frac{1}{4}+\frac{1}{6}-\frac{1}{8}+\cdots$$
$$=\frac{1}{2}\left(1-\frac{1}{2}+\frac{1}{3}-\frac{1}{4}+\cdots\right)=\frac{1}{2}\log 2.$$

Thus, $\log 2=\frac{1}{2}\log 2$.

Comments regarding FFF#28 appear in [CMJ, 23:1, p. 38] and [CMJ, 24:3, p. 231].

FFF #76 Telescoping Series [CMJ, 25:4, p. 309]

Stanley Wertheimer of Connecticut College in New London, CT, has sent in a paper written by a freshman student, Eleanor A. Maddock, that treats the sum

$$\sum_{k=1}^{\infty}\frac{1}{(k+1)(k+2)}.$$

On the one hand,

$$\frac{1}{6}+\frac{1}{12}+\frac{1}{20}+\cdots=\left(\frac{1}{2}-\frac{1}{3}\right)+\left(\frac{1}{3}-\frac{1}{4}\right)+\left(\frac{1}{4}-\frac{1}{5}\right)+\cdots=\frac{1}{2};$$

while on the other,

$$\frac{1}{6}+\frac{1}{12}+\frac{1}{20}+\cdots=\left(1-\frac{5}{6}\right)+\left(\frac{5}{6}-\frac{3}{4}\right)+\left(\frac{3}{4}-\frac{7}{10}\right)+\left(\frac{7}{10}-\frac{2}{3}\right)+\cdots=1.$$

She analyzed this discrepancy and showed where the fault lay, a task that some of your own undergraduates might undertake.

FFF #83 Power Series Thinning [CMJ, 26:1, p. 35]

David Rose of Oral Roberts University in Tulsa, OK, mentions that it is not well known that a function expressed as a power series can be reduced proportionately by deleting the proper proportion of terms from the series. For example,

$$\frac{1}{2}e^x=\sum_{n=0}^{\infty}\frac{x^n}{n!}-\left(1+x+\frac{x^4}{4!}+\frac{x^5}{5!}+\frac{x^8}{8!}+\frac{x^9}{9!}+\cdots\right)$$

is observed by solving the differential equation $y''+y=e^x$ in two ways. The standard method yields the general solution $y(x)=a\cos x+b\sin x+(1/2)e^x$, whereas the series method of solution produces

$$y(x) = a\left(1 - \frac{x^2}{2!} + \frac{x^4}{4!} + \cdots\right) + b\left(x - \frac{x^3}{3!} + \frac{x^5}{5!} - \frac{x^7}{7!} + \cdots\right)$$
$$+ \left(\frac{x^2}{2!} + \frac{x^3}{3!} + \frac{x^6}{6!} + \frac{x^7}{7!} + \cdots\right)$$
$$= a\cos x + b\sin x + \left(\frac{x^2}{2!} + \frac{x^3}{3!} + \frac{x^6}{6!} + \frac{x^7}{7!} + \frac{x^{10}}{10!} + \frac{x^{11}}{11!} + \cdots\right).$$

Subtraction of $a\cos x + b\sin x$ from both solutions yields the result.

FFF #83 Power Series Thinning [CMJ, 26:5, p. 384]

When the differential equation $y'' + y = e^x$ is solved by the standard method, the general solution is $y(x) = a\cos x + b\sin x + (1/2)e^x$, whereas using the series method, we find that

$$y(x) = a\cos x + b\sin x + \sum\left\{\frac{x^k}{k!} : k \not\equiv 0, 1 (\text{mod } 4)\right\}.$$

D. Rose, the contributor of this item [CMJ, 26:1, p. 35], makes the following observations:

> Of course, the constants a and b are not the same in both solutions. In the series solution, a and b are the first two coefficients, i.e., $a = y(0)$ and $b = y'(0)$. The method of solution of $y'' + y = e^x$ by Laplace transform with $y(0) = a$ and $y'(0)$ reveals that
>
> $$y(x) = \left(a - \frac{1}{2}\right)\cos x + \left(b - \frac{1}{2}\right)\sin x + \frac{e^x}{2},$$
>
> so that, actually,
>
> $$\left(\frac{x^2}{2!} + \frac{x^3}{3!} + \frac{x^6}{6!} + \frac{x^7}{7!} + \frac{x^{10}}{10!} + \frac{x^{11}}{11!} + \cdots\right) = \frac{1}{2}(e^x) - \frac{1}{2}(\cos x + \sin x),$$
>
> from which it follows that the flimflam equation is true frequently but sparsely. However, if x is considered a uniform random variable on any interval $[\mu - \pi, \mu + \pi]$ of length 2π, the expected value of the left member of the equation is equal to the expected value of $\frac{1}{2}e^x$ over the interval since the expected value of the error term $\frac{1}{2}(\cos x + \sin x)$ agrees with the average value of 0. Thus, both $1 + x + x^4/4! + x^5/5! + x^8/8! + x^9/9! + \cdots$ and $x^2/2! + x^3/3! + x^6/6! + x^7/7! + x^{10}/10! + \cdots$ though different, are unbiased estimators of $(e^\mu \sinh \pi)/2\pi$, the expected value of $e^x/2$ on the interval $[\mu - \pi, \mu + \pi]$.

FFF #135 Positive Series with a Negative Sum [CMJ, 29:5, p. 407]

William A. Simpson from Michigan State University in East Lansing submits some analysis that would do an eighteenth-century mathematician proud. Consider this infinite matrix:

$$\begin{pmatrix} 1 & 1 & 1 & 1 & \cdots \\ 1 & \frac{1}{2} & \frac{1}{4} & \frac{1}{8} & \cdots \\ 1 & \frac{1}{3} & \frac{1}{9} & \frac{1}{27} & \cdots \\ 1 & \frac{1}{4} & \frac{1}{16} & \frac{1}{64} & \cdots \\ & & \cdots & & \end{pmatrix}$$

Summing up the row totals yields

$$\begin{aligned}(1+1+\cdots)+(1+1)+\left(1+\frac{1}{2}\right)+\left(1+\frac{1}{3}\right)+\cdots \\ =1+1+1+\cdots+1+\frac{1}{2}+\frac{1}{3}+\cdots \\ =\sum_{k=1}^{\infty}1+\sum_{k=1}^{\infty}\frac{1}{k},\end{aligned}$$

while summing over the column totals yields

$$\sum_{k=1}^{\infty}1+\sum_{k=1}^{\infty}\frac{1}{k}+\sum_{k=1}^{\infty}\frac{1}{k^2}+\sum_{k=1}^{\infty}\frac{1}{k^3}+\cdots.$$

Since both of these sum the terms in the matrix, they should be equal. Canceling common terms yields

$$\begin{aligned}0&=\sum\frac{1}{k^2}+\sum\frac{1}{k^3}+\sum\frac{1}{k^4}+\sum\frac{1}{k^5}+\cdots \\ &=\frac{\pi^2}{6}+\sum\frac{1}{k^3}+\frac{\pi^4}{90}+\sum\frac{1}{k^5}+\cdots,\end{aligned}$$

so that $\sum(1/k^3)+\sum(1/k^5)+\cdots$ must be negative.

FFF #141 Evaluation of a sum

[CMJ, 30:2, pp. 130–131]

Problem. Evaluate

$$\sum_{n=1}^{\infty}\frac{1}{(2n-1)(3n-1)}.$$

Solution by Joe Howard, New Mexico Highlands University, Las Vegas, NM.

$$\begin{aligned}\sum_{n=1}^{\infty}\frac{1}{(2n-1)(3n-1)}&=\sum_{n=1}^{\infty}\left(\frac{2}{2n-1}-\frac{3}{3n-1}\right) \\ &=\frac{2}{1}-\frac{3}{2}+\frac{2}{3}-\frac{3}{5}+\frac{2}{5}-\frac{3}{8}+\cdots.\end{aligned}$$

Let

$$f(x) = \frac{2x}{1} - \frac{3x^2}{2} + \frac{2x^3}{3} - \frac{3x^5}{5} + \cdots$$
$$= 2\left(\frac{x}{1} + \frac{x^3}{3} + \frac{x^5}{5} + \cdots\right) - 3\left(\frac{x^2}{2} + \frac{x^5}{5} + \frac{x^8}{8} + \cdots\right)$$

for $0 \le x \le 1$. The series is absolutely convergent for $|x| < 1$, so the terms can be rearranged. It follows that

$$f'(x) = 2(1 + x^2 + x^4 + \cdots) - 3(x + x^4 + x^7 + \cdots)$$
$$= 2\left(\frac{1}{1-x^2}\right) - 3x\left(\frac{1}{1-x^3}\right) = \frac{x+2}{(x+1)(x^2+x+1)}$$
$$= \frac{1}{x+1} + \frac{-x+1}{x^2+x+1}.$$

Thus,

$$f(x) = \int_0^x \frac{dt}{t+1} - \frac{1}{2}\int_0^x \frac{(2t+1)dt}{t^2+t+1} + \frac{3}{2}\int_0^x \frac{dt}{\left(t+\frac{1}{2}\right)^2 + \left(\frac{\sqrt{3}}{2}\right)^2}$$
$$= \log|x+1| - \frac{1}{2}\log|x^2+x+1| + \sqrt{3}\left(\arctan\frac{2x+1}{\sqrt{3}} - \arctan\frac{1}{\sqrt{3}}\right).$$

Since the series for $f(x)$ is convergent for $x = 1$, the given series converges to

$$f(1) = \log\frac{2}{\sqrt{3}} + \frac{\sqrt{3}\cdot\pi}{6}.$$

This problem was posed in *Crux Mathematicorum* 22 (1996), 219; 23 (1997), 371–373, where the (correct) answer given was $2\log 2 - \frac{3}{2}\log 3 + \frac{\sqrt{3}\cdot\pi}{6}$.

FFF #206 A series that converges and diverges [CMJ, 34:2, p. 135]

Consider the infinite series

$$\sum_{n=2}^{\infty}(-1)^n \ln\left(1 - \frac{1}{n}\right).$$

This series can be shown to converge by the alternating series test. But we also have

$$\sum_{n=2}^{\infty}(-1)^n \ln\left(1 - \frac{1}{n}\right)$$
$$= \sum_{n=2}^{\infty}(-1)^n \ln\left(\frac{n-1}{n}\right)$$

$$
\begin{aligned}
&= \sum_{n=2}^{\infty}(-1)^n[\ln(n-1)-\ln n] \\
&= [\ln 1 - \ln 2] - [\ln 2 - \ln 3] + [\ln 3 - \ln 4] - [\ln 4 - \ln 5] + \cdots \\
&= -2\ln 2 + 2\ln 3 - 2\ln 4 + 2\ln 5 - 2\ln 6 + \cdots \\
&= \sum_{n=2}^{\infty}(-1)^{n+1}(2\ln n).
\end{aligned}
$$

This last series diverges since the nth term does not tend to zero.

FFF #218 Alternating madness

[CMJ, 35:1, pp. 40–41]

Taking note of the manipulation of an alternating series to a divergent series on **FFF #206**, Ollie Nanyes of Bradley University in Peoria, IL sends along some shenanigans with a more mundane alternating series.

First, we have that

$$
\begin{aligned}
\sum_{n=1}^{\infty}(-1)^{n+1}\frac{1}{n} &= 1 - \frac{1}{2} + \frac{1}{3} - \frac{1}{4} + \frac{1}{5} - \frac{1}{6} + \frac{1}{7} - \cdots \\
&= 1 + \left(\frac{1}{2} - 1\right) + \left(1 - \frac{2}{3}\right) + \left(\frac{3}{4} - 1\right) + \cdots \\
&\quad + \left(\frac{2n-1}{2n} - 1\right) + \left(1 - \frac{2n}{2n+1}\right) + \cdots \\
&= 1 + \frac{1}{2} + (-1+1) - \frac{2}{3} + \frac{3}{4} + (-1+1) - \cdots \\
&\quad + \frac{2n-1}{2n} + (-1+1) - \frac{2n}{2n+1} - \cdots \\
&= 1 + \left(\frac{1}{2} - \frac{2}{3}\right) + \left(\frac{3}{4} - \frac{4}{5}\right) + \cdots + \left(\frac{2n-1}{2n} - \frac{2n}{2n+1}\right) + \cdots \\
&= 1 + \left(-\frac{1}{6}\right) + \left(-\frac{1}{20}\right) + \cdots + \left(\frac{4n^2 - 1 - 4n^2}{4n^2 + 2n}\right) + \cdots \\
&= 1 - \sum_{n=1}^{\infty}\frac{1}{2n(2n+1)},
\end{aligned}
$$

which is absolutely convergent.

On the other hand,

$$
\begin{aligned}
\sum_{n=1}^{\infty}(-1)^{n+1}\frac{1}{n} &= 1 - \frac{1}{2} + \frac{1}{3} - \frac{1}{4} + \frac{1}{5} - \frac{1}{6} + \frac{1}{7} - \cdots \\
&= 1 + \left(\frac{1}{2} - 1\right) + \frac{1}{3} + \left(\frac{3}{4} - 1\right) + \frac{1}{5} + \cdots
\end{aligned}
$$

$$
\begin{aligned}
&+\left(\frac{2n-1}{2n}-1\right)+\frac{1}{2n+1}+\cdots \\
&=1+\frac{1}{2}+\left(-1+\frac{1}{3}\right)+\frac{3}{4}+\left(-1+\frac{1}{5}\right)+\frac{5}{6}+\cdots \\
&\quad+\frac{2n-1}{2n}+\left(-1+\frac{1}{2n+1}\right)+\cdots \\
&=1+\frac{1}{2}-\frac{2}{3}+\frac{3}{4}-\frac{4}{5}+\cdots+\frac{2n-1}{2n}-\frac{2n}{2n+1}+\cdots \\
&=1+\sum_{n=2}^{\infty}(-1)^n\frac{n-1}{n},
\end{aligned}
$$

which is divergent.

More generally, Nanyes points out that any alternating series $\sum_{n=1}^{\infty}(-1)^n a_n$, with $a_n > 0$, can be manipulated to a divergent series $\sum_{n=1}^{\infty}(-1)^{n+1}(1-a_n)$ by replacing each oddly indexed term $-a_{2k+1}$ by $(1-a_{2k+1})-1$ and rebracketing.

6.5 Answers to Puzzles

Solution to Puzzle 1. In both cases the answer is no. To see this, assign a numerical value to each square on the grid in the following manner. In any given moment, we say that the value of the configuration is the total sum of the values of all the occupied squares. The reason for this numbering is as follows: when a stone is removed from a particular square, it is replaced by two stones on squares that each have half the value. That is, the total value of the configuration is *not* changed throughout the solving process.

To settle the first question, we need to know the total value of all the squares on the grid. To do this, we must evaluate an infinite series, indeed a multiple series. (This is where infinite series come into play in this problem ... after reading this solution, notice

⋮		⋮		⋮		⋰
$\frac{1}{32}$	$\frac{1}{64}$	$\frac{1}{128}$	$\frac{1}{256}$	$\frac{1}{512}$	$\frac{1}{1024}$	…
$\frac{1}{16}$	$\frac{1}{32}$	$\frac{1}{64}$	$\frac{1}{128}$	$\frac{1}{256}$	$\frac{1}{512}$	
$\frac{1}{8}$	$\frac{1}{16}$	$\frac{1}{32}$	$\frac{1}{64}$	$\frac{1}{128}$	$\frac{1}{256}$	
$\frac{1}{4}$	$\frac{1}{8}$	$\frac{1}{16}$	$\frac{1}{32}$	$\frac{1}{64}$	$\frac{1}{128}$	…
$\frac{1}{2}$	$\frac{1}{4}$	$\frac{1}{8}$	$\frac{1}{16}$	$\frac{1}{32}$	$\frac{1}{64}$	
1	$\frac{1}{2}$	$\frac{1}{4}$	$\frac{1}{8}$	$\frac{1}{16}$	$\frac{1}{32}$	…

FIGURE 6.4
Labeling the Quarter-Checkerboard

that this problem would be much harder without the concept of infinite series. They are the vehicle by which we can consider the infinite grid as *a finite resource*.) Since all the terms are positive, we can freely use rearrangement and scrambling. The first row has value $1+1/2+1/4+\cdots = 2$, the second row has value $1/2+1/4+1/8+\cdots = 1$. Similarly, the third row has value $1/2$, the fourth has value $1/4$, etc. Summing now the rows, the grid has total value $2+1+1/2+1/4+1/8+\cdots = 4$. Now, the six stones in the colored areas have a total value of $1+1/2+1/2+1/4+1/4+1/4 = 2\frac{3}{4}$, and therefore the total remaining part of the board can only ever be worth the remaining $1\frac{1}{4}$. Since the configuration will always have value $2\frac{3}{4} > 1\frac{1}{4}$ (by the previous paragraph), it cannot fit into the uncolored part of the grid.

The second question is slightly more difficult to analyze. In this case, the value of the starting configuration is 1, and there is still a total of $1\frac{1}{4}$ available outside the colored squares. However, observe that we cannot actually use all the uncolored squares. First, it is clear that there will always be exactly one stone in the first column. (Stones only move up and to the right, so that no new stones can enter the first column from other columns; whenever the stone currently in the first column is touched, exactly one new stone is placed in the first column.) Now, the total "available" area in the first column is $1/8 + 1/16 + 1/32 + \cdots = 1/4$. However, we can use at most $1/8$ from that total, so at least $1/8$ will be wasted. Similarly, $1/8$ will be wasted from the bottom row. Having eliminated now $1/4$, our maximum possible area is now exactly 1. We can now reach the desired total if and only if all other squares are occupied; this includes of course an infinite number of squares. However, if the puzzle is to be solved, it must be solved after a finite number of steps. Since then there will be only a finite number of stones on the board, the task is impossible.

Solution to Puzzle 2. This is a trick question. In fact, it is impossible to get a checker to the fifth row above the line, so that $c_5, c_6, c_7, \ldots$ mean nothing!

Suppose for the sake of contradiction that some strategy will place a checker in the fifth row. As in the solution to the previous problem, we solve this by assigning numerical values to the board. Suppose for simplicity that the checker in the fifth row is in the "center" column. Now, we assign values to all the squares in the grid according to the diagram. That is, the values of the squares decrease by a factor of $r = (-1+\sqrt{5})/2$ with each step

1	r^{-1}	r^{-2}	r^{-3}	r^{-2}	r^{-1}	1
r	1	r^{-1}	r^{-2}	r^{-1}	1	r
r^2	r	1	r^{-1}	1	r	r^2
r^3	r^2	r	1	r	r^2	r^3
r^4	r^3	r^2	r	r^2	r^3	r^4
r^5	r^4	r^3	r^2	r^3	r^4	r^5

FIGURE 6.5
Labeling the Infinite Checkerboard

downward or away from the center column. Why have we chosen this value of r? It is the positive solution of the equation $r^2 + r = 1$. Therefore, an upward jump or a jump toward the center removes two checkers and creates a checker with the same total value as the removed checkers. Other sorts of jumps will remove two checkers and create a checker of lesser value. As in the previous problem, we evaluate the total value of the bottom section of the grid. The center column below the line has value $1+r+r^2+\cdots+=\frac{1}{1-r}$. Similarly, we see that the two columns n spaces away from the center each have total value $\frac{r^n}{1-r}$. Then we have a total value of

$$\frac{1}{1-r}+2\left(\frac{r}{1-r}+\frac{r^2}{1-r}+\frac{r^3}{1-r}+\cdots\right)=\frac{1}{1-r}+\frac{2r}{(1-r)^2}=\frac{1+r}{(1-r)^2}.$$

We begin with a finite collection of checkers, so the starting value of the configuration is *less* than $(1+r)/(1-r)^2$. As jumps are taken, this value can only decrease. Now, the value of the single checker in the fifth row is r^{-5}, so we must have

$$r^{-5} < \frac{1+r}{(1-r)^2}.$$

However, a moment of algebra (or a moment with a calculator) shows that both sides of the above inequality have the common value $11.0901699\ldots$, contradicting the inequality.

Solution to Puzzle 3. We claim that, in n trips with the jeep, Desert Dan can actually travel $25H_n$ miles, where H_n is the nth harmonic number. (We really hope the reader finds this last appearance of the harmonic series surprising. We said that it shows up frequently.) Since the harmonic series has unbounded partial sums, this would imply that Desert Dan could eventually cross any arbitrarily large desert. For convenience, we will measure fuel in *miles*, so that 5 miles of fuel is the amount of fuel that would carry the jeep 5 miles.

In describing Desert Dan's strategy, we will number his trips in reverse. That is, his chronologically first trip is called trip n, his next trip is trip $n-1$, and so on until trip 1, which happens last. The strategy operates as follows. On trip n, Desert Dan drives $25/n$ miles. He will need $25/n$ miles of fuel to return, so he leaves the remaining $50(n-1)/n$ miles in a container. Call this spot Checkpoint n. On trip $n-1$, Desert Dan drives to Checkpoint n and uses the fuel there to fill his tank totally. He then travels $25/(n-1)$ miles, and leaves $50(n-2)/(n-1)$ miles of fuel at Checkpoint $n-1$. He now has exactly $25/(n-1)$ miles of fuel, which is just enough to get him back to Checkpoint n, where he takes $25/n$ miles of fuel from the cache to get himself home. The general strategy for trip k is as follows: drive to all the already-established checkpoints, filling up on fuel at every checkpoint. From the last checkpoint, drive $25/k$ miles into the desert, leaving $25(k-1)/k$ miles of fuel at the newly established checkpoint k. Using the last of your fuel, return to the last checkpoint. On the return trip, take from each cache just enough fuel to reach the next checkpoint. Now we must verify that this strategy works. We first observe that the distance from Checkpoint i to Checkpoint $i+1$ is $25/i$ miles (where Checkpoint $n+1$ is the edge of the desert), so that indeed Checkpoint 1 is the promised $25H_n$ miles from the edge of the desert. But will the caches hold enough fuel for Desert Dan to accomplish his mission? In our strategy, the cache at Checkpoint i has $50(i-1)/i$ miles. This fuel must supply each of $i-1$ trips after trip i (remember the reverse numbering of the trips). Each such trip will use $25/i$ miles of fuel to replenish after the trip from Checkpoint $i+1$ and another $25/i$ to make the return trip to Checkpoint $i+1$. Thus each of $i-1$ trips will draw $50/i$ miles of fuel from the cache. We have $50(i-1)/i$ miles of fuel, so the fuel will last.

A

101 True or False Questions

The authors of this book believe that true or false questions[1] have an educational value, particularly in learning mathematics. Sorting between true and false propositions forces the reader to understand clearly the relationships between hypotheses and conclusions in theorems and to carefully acknowledge and avoid misconceptions and faulty generalizations. Discerning the difference between a true-seeming statement and a genuinely true statement signifies a maturity of understanding of the fundamentals of a mathematical theory.

In keeping with these beliefs, we present the reader with a collection of true or false statements on the subject of infinite series. There are at least two distinct reasons why the reader may find these useful. Most obviously, the reader is encouraged to read through the questions and try to answer each for himself or herself. For the reader's convenience in this endeavor, we have included each question twice, first in a section containing only the questions and again in a section with answers and short explanations. This appendix should not be interpreted as a "comprehensive final exam" of this text in any sense. No effort was made to ensure that every bit of content in the chapters of this book found its way into these questions. On the other hand, the nuances of these true or false questions can be seen as a test of the reader's reasoning ability, which we hope that reading this book has developed.

The other use is for instructors who are teaching the subject of infinite series to their students, perhaps as part of a calculus or analysis course. We encourage you to consider incorporating true and false questions in your assessment, and we invite you to select any questions from this chapter that you feel suit the difficulty level and focus of your course. As always we do not intend this collection to be exhaustive, and we have no doubt that the reader will be able to devise more questions more suitable to his or her circumstances.

[1]The pedantic reader may wonder why we call these objects questions even though they are grammatically statements and have no question mark. It seems inappropriate to call them statements because that suggests an assertion, and of course we make no claim that the 101 statements in this appendix are true. The "questions" are implicit, however, because each statement is to be interpreted as "Is it true or false that ... ?"

Questions

T/F 1 *The series* $\sum_{n=1}^{\infty} a_n$ *converges whenever* $\lim_{n\to\infty} a_n = 0$.

T/F 2 *The series* $\sum_{n=1}^{\infty} a_n$ *converges whenever the sequence of partial sums converges to* 0.

T/F 3 *If* $\sum_{n=1}^{\infty} a_n$ *converges and* $a_n > 0$, *then* $\sum_{n=1}^{\infty} a_n^2$ *converges.*

T/F 4 *If* $\sum_{n=1}^{\infty} a_n$ *converges and* $a_n > 0$, *then* $\sum_{n=1}^{\infty} \sqrt{a_n}$ *converges.*

T/F 5 *If* $\sum_{n=1}^{\infty} a_n$ *converges and* $a_n > 0$, *then* $\sum_{n=1}^{\infty} \frac{1+a_n}{2+a_n}$ *converges.*

T/F 6 *If* $\sum_{n=1}^{\infty} a_n$ *converges and* $a_n > 0$, *then* $\sum_{n=1}^{\infty} \frac{2^n+a_n}{3^n+a_n}$ *converges.*

T/F 7 $\sum_{n=1}^{\infty} \sin \frac{1}{n^2}$ *converges.*

T/F 8 $\sum_{n=1}^{\infty} \cos \frac{1}{n^2}$ *converges.*

T/F 9 $\sum_{n=1}^{\infty} \frac{1}{n^{1+1/n}}$ *converges.*

T/F 10 *If* p *is a fixed real number, then* $\sum_{n=1}^{\infty} 1/n^p$ *converges if and only if* $p > 1$.

T/F 11 *If* $\sum_{n=1}^{\infty} a_n$ *converges, then* $\sum_{n=1}^{\infty} \frac{a_n}{n}$ *converges.*

T/F 12 *The series* $\sum_{n=1}^{\infty} n^{\cos 3}$ *converges.*

T/F 13 $\sum_{n=1}^{\infty} \frac{n+1}{n+5}$ *converges to 1/5.*

T/F 14 $\sum_{n=2}^{\infty} 3(4)^{-n+1}$ *converges to 1.*

T/F 15 *If* $a_n > 0$ *and the sequence of partial sums for* $\sum_{n=1}^{\infty} a_n$ *is bounded above, then the series converges.*

T/F 16 *If the sequence of partial sums for* $\sum_{n=1}^{\infty} a_n$ *is bounded, then the series converges.*

T/F 17 *Every rearrangement of an absolutely convergent series has the same sum.*

T/F 18 *If the Strengthened Root Test indicates absolute convergence for a given series, then the Strengthened Ratio Test will also indicate absolute convergence for that series.*

T/F 19 *If* $\lim_{n\to\infty} a_n/b_n = \infty$ *and* $\sum_{n=1}^{\infty} b_n$ *diverges, then* $\sum_{n=1}^{\infty} a_n$ *diverges.*

T/F 20 *If* $\lim_{n\to\infty} a_n/b_n = 0$ *and* $\sum_{n=1}^{\infty} b_n$ *converges, then* $\sum_{n=1}^{\infty} a_n$ *converges.*

T/F 21 *If* $\sum_{n=1}^{\infty} a_n = A$ *and* $\sum_{n=1}^{\infty} |a_n| = B$ *and* A *and* B *are finite, then* $|A| = B$.

T/F 22 *If* $\sum_{n=1}^{\infty} a_n$ *diverges and* $a_n > 0$, *then* $\sum_{n=1}^{\infty} \frac{a_n}{1+a_n}$ *diverges also.*

T/F 23 *If both* $\sum_{n=1}^{\infty} a_n$ *and* $\sum_{n=1}^{\infty} b_n$ *diverge, then* $\sum_{n=1}^{\infty}(a_n + b_n)$ *diverges.*

T/F 24 *If both* $\sum_{n=1}^{\infty} a_n$ *and* $\sum_{n=1}^{\infty} b_n$ *diverge, then* $\sum_{n=1}^{\infty}(a_n + b_n)$ *converges.*

T/F 25 *If* $\sum_{n=1}^{\infty} a_n$ *converges and* $\sum_{n=1}^{\infty} b_n$ *diverges, then* $\sum_{n=1}^{\infty}(a_n + b_n)$ *diverges.*

T/F 26 *If* $\sum_{n=1}^{\infty} a_n$ *converges and* $\sum_{n=1}^{\infty} b_n$ *diverges, then* $\sum_{n=1}^{\infty}(a_n + b_n)$ *converges.*

T/F 27 *If* $a_n > 0$ *and* $\sum_{n=1}^{\infty} a_n$ *converges, then* $\sum_{n=1}^{\infty} 1/a_n$ *diverges.*

T/F 28 *If* $a_n > 0$ *and* $\sum_{n=1}^{\infty} a_n$ *diverges, then* $\sum_{n=1}^{\infty} 1/a_n$ *converges.*

T/F 29 *If* $\lim_{n\to\infty} a_n/b_n = 1$ *and* $\sum_{n=1}^{\infty} b_n$ *converges, then* $\sum_{n=1}^{\infty} a_n$ *converges.*

T/F 30 *If* $\lim_{n\to\infty} a_n = 0$ *and* $\sum_{n=1}^{\infty} b_n$ *converges, then* $\sum_{n=1}^{\infty} a_n b_n$ *converges.*

T/F 31 *If* $\lim_{n\to\infty} a_n = 0$, $0 < a_{n+1} \leq a_n$ *and* $\sum_{n=1}^{\infty} b_n$ *converges, then* $\sum_{n=1}^{\infty} a_n b_n$ *converges.*

T/F 32 *If* $\lim_{n\to\infty} a_n = 0$ *and* $\sum_{n=1}^{\infty} b_n$ *converges absolutely, then* $\sum_{n=1}^{\infty} a_n b_n$ *converges.*

T/F 33 *If* $a_n \geq 0$, *then* $\sum_{n=1}^{\infty} a_n$ *does not converge conditionally.*

T/F 34 *The Alternating Series Test is a test for conditional convergence.*

T/F 35 *If* $\sum_{n=1}^{\infty} a_n b_n$ *converges, then* $\sum_{n=1}^{\infty} a_n$ *and* $\sum_{n=1}^{\infty} b_n$ *converge.*

T/F 36 *If* $a_n > 0$ *and* $\sum_{n=1}^{\infty} a_n$ *converges, then* $\sum_{n=1}^{\infty} \sin a_n$ *converges.*

T/F 37 *If* $\sum_{n=1}^{\infty} a_n$ *converges, then* $\sum_{n=1}^{\infty} \cos a_n$ *converges.*

T/F 38 *If* $0 < a_{n+1} < a_n$ *for all* n *and* $\lim_{n\to\infty} a_n = 0$, *then*

$$\sum_{n=1}^{\infty}(-1)^{n+1}a_n$$

converges and has sum S *satisfying* $|S_n - S| < a_{n+1}$.

T/F 39

$$\sum_{n=1}^{\infty}\left(1-\frac{1}{n}\right)^{n}$$

is a divergent series.

T/F 40 *The Root Test cannot be used to determine conditional convergence.*

T/F 41 *The Ratio Test cannot be used to determine conditional convergence.*

T/F 42

$$\sum_{n=1}^{\infty}\frac{1}{n^{n}}\geq 2.$$

T/F 43 *The Ratio Test cannot be used to determine the convergence or divergence of the series*

$$\sum_{n=1}^{\infty}\frac{n}{n^{3}+5}.$$

T/F 44 *If $\sum_{n=1}^{\infty} a_n$ converges, then $\sum_{n=1}^{\infty}(-1)^n a_n$ converges also.*

T/F 45

$$\frac{1}{2}+\left(\frac{1}{2}\right)^{2}+\left(\frac{1}{2}\right)^{3}+\cdots+\left(\frac{1}{2}\right)^{10000}<1.$$

T/F 46 *If $\sum_{n=1}^{\infty} b_n$ converges and if $0 \leq a_{n+k} \leq b_n$ for some $k \geq 1000$ and every $n \geq 1000$, then $\sum_{n=1}^{\infty}(-1)^n a_n$ converges absolutely.*

T/F 47

$$\sum_{n=1}^{\infty}\frac{1}{\log(n^{6}+1)}$$

converges.

T/F 48 *If $a_n \leq b_n \leq 0$ and $\sum_{n=1}^{\infty} b_n$ diverges, then $\sum_{n=1}^{\infty} a_n$ diverges also.*

T/F 49 *If $a_n \leq b_n \leq 0$ and $\sum_{n=1}^{\infty} a_n$ converges, then $\sum_{n=1}^{\infty} b_n$ converges also.*

T/F 50 *If $a_n > 0$ and $\sum_{n=1}^{\infty} a_n$ converges, then $\sum_{n=1}^{\infty}(-1)^n a_n$ converges also.*

T/F 51 *If the condition* $0 \le a_{n+1} \le a_n$ *fails (even eventually) when applying the Alternating Series Test, then the series* $\sum_{n=1}^{\infty}(-1)^n a_n$ *diverges.*

T/F 52 *If the condition* $\lim_{n\to\infty} a_n = 0$ *fails when applying the Alternating Series Test, then the series* $\sum_{n=1}^{\infty}(-1)^n a_n$ *diverges.*

T/F 53 *If the Ratio Test indicates absolute convergence for a particular series, then the Root Test will also indicate absolute convergence for that series.*

T/F 54

$$\sum_{n=1}^{\infty} \frac{n+2}{n+4} = \infty.$$

T/F 55 *If* $\sum_{n=1}^{\infty} a_n = 7$ *and* $a_1 = a_2 = 3$, *then* $\sum_{n=3}^{\infty} a_n = 1$.

T/F 56 $3/2, 4/3, 5/4, 6/5, \ldots, (n+1)/n, \ldots$ *is a convergent series.*

T/F 57 *The first three terms in the sequence of partial sums associated with* $\sum_{n=1}^{\infty}(1/2)^n$ *are 1/2,3/4,7/8.*

T/F 58

$$\sum_{n=2}^{\infty} \left(\frac{1}{3n+1} - \frac{1}{3n+4} \right) = \frac{1}{7}.$$

T/F 59 *If* $\lim_{n\to\infty} a_{2n} = 3$ *and* $\lim_{n\to\infty} a_{2n+1} = 3$, *then* $\lim_{n\to\infty} a_n = 3$.

T/F 60

$$\lim_{n\to\infty} a_n = 0 \Leftrightarrow \lim_{n\to\infty} |a_n| = 0.$$

T/F 61 *If* $0 \le a_n \le b_n$ *and* $\sum_{n=1}^{\infty} b_n$ *diverges, then* $\sum_{n=1}^{\infty} a_n$ *diverges also.*

T/F 62 *If a series converges, then no subsequence of the sequence of partial sums can be unbounded.*

T/F 63 *The Integral Test is of no value in testing* $\sum_{n=1}^{\infty} \sin n/n$ *for convergence or divergence.*

T/F 64 *The sum of the infinite series* $\sum_{n=1}^{\infty} a_n$ *is the limit*

$$S = \lim_{k\to\infty} \sum_{n=1}^{k} a_n,$$

whenever this limit exists.

T/F 65 $\sum_{n=1}^{\infty} a_n$ *converges if and only if* $\sum_{n=p}^{\infty} a_n$ *converges for every* p.

T/F 66 *The Ratio Test is of no value in testing* $\sum_{n=1}^{\infty} 1/\sqrt{n(n+1)(n+2)}$ *for convergence or divergence.*

T/F 67

$$\sum_{n=2}^{\infty}\left(1-\frac{1}{n^2}\right)^{n^2}$$

diverges.

T/F 68 *If* $a_n > 0$ *for all* n *and if* $\sum_{n=1}^{\infty} a_n$ *converges, then*

$$\lim_{n\to\infty}\frac{a_{n+1}}{a_n} = L < 1.$$

T/F 69 *If for some constant* c *we have* $ca_n \geq 1/n$, *then necessarily* $\sum_{n=1}^{\infty} a_n$ *diverges.*

T/F 70 *The series* $\sum_{n=2}^{\infty}(-1)^n$ *has as its sequence of partial sums* $1, 0, 1, 0, \ldots$

T/F 71 *If* $\lim_{n\to\infty} a_n = 0$, *then*

$$\sum_{n=1}^{\infty}\frac{a_n+2}{a_n^2+1} = 2.$$

T/F 72 *If* $\lim_{n\to\infty} a_n = 0$, *then*

$$\sum_{n=1}^{\infty}\frac{a_n}{a_n^2+n^2}$$

converges absolutely.

T/F 73

$$\sum_{n=1}^{\infty} n\left(\frac{1}{2}\right)^n = 2.$$

T/F 74

$$\sum_{k=1}^{\infty}\left(\text{Arctan}\,\frac{1}{k} - \text{Arctan}\,\frac{1}{k+1}\right) = \frac{\pi}{4}.$$

T/F 75 *The Ratio Test is of no help in determining whether*

$$\sum_{n=1}^{\infty} \frac{(2n)!}{(n!)^2}.$$

converges or diverges.

T/F 76 *If* $|a_{n+1}/a_n| > 1$ *for all n, then* $\sum_{n=1}^{\infty} |a_n|$ *diverges.*

T/F 77 *If* $|a_{n+1}/a_n| < 1$ *for all n, then* $\sum_{n=1}^{\infty} |a_n|$ *converges.*

T/F 78 *If* $s_n = 3 + 2^{-n}$, *then* $\lim_{n\to\infty} s_n = 4$.

T/F 79 *The first three terms in the sequence of partial sums associated to the series* $\sum_{n=1}^{\infty}(-1/2)^n$ *are* $-1/2, 1/4, -1/8$.

T/F 80

$$\sum_{n=1}^{\infty} 3^{-n} 2^{n+1} = 4.$$

T/F 81 *Every partial sum of*

$$\sum_{n=1}^{\infty} \left(\frac{1}{n(n+1)}\right)$$

is less than 1.

T/F 82 *The Integral Test can be used to show that*

$$\sum_{n=3}^{\infty} \frac{(-1)^n}{n \log n \log(\log n)}$$

does not converge absolutely.

T/F 83 *The insertion or removal of a finite number of terms in a series cannot affect its convergence or divergence, although it may affect its sum.*

T/F 84 *If a series converges, then any summation of a subset of the terms will form a convergent series.*

T/F 85 *If* $\sum_{n=1}^{\infty} a_n$ *converges and* $\{b_n\}_{n=1}^{\infty}$ *is a bounded sequence, then* $\sum_{n=1}^{\infty} a_n b_n$ *converges.*

T/F 86 *If* $\sum_{n=1}^{\infty} a_n$ *diverges, then* $\sum_{n=1}^{\infty} |a_n|$ *must diverge.*

T/F 87 *If* $a_n = 3/n^2$, *then* $\sum_{n=1}^{\infty}(a_n - a_{n+1})$ *converges to zero.*

T/F 88 *A series representation for the decimal* 0.999999... *can be given by*

$$\sum_{n=1}^{\infty} 9\left(\frac{1}{10}\right)^n$$

where the sum is exactly 1.0, *so* $0.999999999999\ldots = 1$.

T/F 89 *If* $\lim_{n\to\infty} a_n$ *does not exist, then we can always correctly claim that* $\sum_{n=1}^{\infty} a_n$ *diverges.*

T/F 90 *A convergent series is either absolutely convergent or conditionally convergent.*

T/F 91 *The alternating harmonic series cannot be rearranged to diverge to* $-\infty$.

T/F 92 *If* $a_n > 0$ *and* $\sqrt[n]{a_n} < 1$ *for all n, then* $\sum_{n=1}^{\infty} a_n$ *converges.*

T/F 93 *If* $a_n > 0$ *and* $\sqrt[n]{a_n} \geq 1$ *for all n, then* $\sum_{n=1}^{\infty} a_n$ *diverges.*

T/F 94 *If* $a_n > 0$ *and* $\sqrt[n]{a_n} < r < 1$ *for all n and a constant r, then* $\sum_{n=1}^{\infty} a_n$ *converges.*

T/F 95 *If* $\lim_{n\to\infty} |a_{n+1}/a_n| = L < 1$ *for some constant L, then* $\lim_{n\to\infty} a_n = 0$.

T/F 96 *If* $\sum_{n=1}^{\infty} a_n^2$ *converges, then* $\sum_{n=1}^{\infty} \frac{a_n}{n}$ *converges.*

T/F 97 *If* $a_n \geq 0$ *and* $\sum_{n=1}^{\infty} a_n$ *converges, then* $\sum_{n=1}^{\infty} \frac{\sqrt{a_n}}{n^p}$ *converges for* $p > 1/2$.

T/F 98 *If* $a_n \geq 0$ *and* $\sum_{n=1}^{\infty} a_n$ *diverges, then* $\sum_{n=1}^{\infty} \frac{a_n}{1+na_n}$ *diverges.*

T/F 99 *If* $a_n \geq 0$ *and* $\sum_{n=1}^{\infty} a_n$ *diverges, then* $\sum_{n=1}^{\infty} \frac{a_n}{1+na_n}$ *converges.*

T/F 100 *If* $a_n \geq 0$ *and* $\sum_{n=1}^{\infty} a_n$ *diverges, then* $\sum_{n=1}^{\infty} \frac{a_n}{1+a_n^2}$ *diverges.*

T/F 101 *If* $a_n \geq 0$ *and* $\sum_{n=1}^{\infty} a_n$ *diverges, then* $\sum_{n=1}^{\infty} \frac{a_n}{1+a_n^2}$ *converges.*

Explanations

T/F 1 *The series* $\sum_{n=1}^{\infty} a_n$ *converges whenever* $\lim_{n\to\infty} a_n = 0$.

Solution. False. The Divergence Test (Corollary 1.44.1) guarantees that the series can converge only if the terms tend to 0. However, the converse fails (the harmonic series, for example, is a counterexample).

T/F 2 *The series $\sum_{n=1}^{\infty} a_n$ converges whenever the sequence of partial sums converges to* 0.

Solution. True. By Definition 1.34, a series converges if and only if its sequence of partial sums converge to *any* number, and zero is acceptable.

T/F 3 *If $\sum_{n=1}^{\infty} a_n$ converges and $a_n > 0$, then $\sum_{n=1}^{\infty} a_n^2$ converges.*

Solution. True. By Theorem 1.44, we have $a_n \to 0$ and consequently $a_n < 1$ eventually. Then also $a_n^2 < a_n$ eventually, so that the Comparison Test applies.

T/F 4 *If $\sum_{n=1}^{\infty} a_n$ converges and $a_n > 0$, then $\sum_{n=1}^{\infty} \sqrt{a_n}$ converges.*

Solution. False. Consider $a_n = 1/n^2$, $\sqrt{a_n} = 1/n$. The former converges and the latter diverges by Theorem 1.59.

T/F 5 *If $\sum_{n=1}^{\infty} a_n$ converges and $a_n > 0$, then $\sum_{n=1}^{\infty} \frac{1+a_n}{2+a_n}$ converges.*

Solution. False. By Theorem 1.44 we have, $a_n \to 0$ and thus $\frac{1+a_n}{2+a_n} \to 1/2$. Thus the latter series' terms do not go to 0, hence it diverges.

T/F 6 *If $\sum_{n=1}^{\infty} a_n$ converges and $a_n > 0$, then $\sum_{n=1}^{\infty} \frac{2^n+a_n}{3^n+a_n}$ converges.*

Solution. True. By Theorem 1.44 we have, $a_n \to 0$. Now apply the ratio test.

$$\frac{a_{n+1}}{a_n} = \frac{2^{n+1} + a_{n+1}}{2^n + a_n} \cdot \frac{3^n + a_n}{3^{n+1} + a_{n+1}}.$$

For large n, the a_i are small and this ratio tends then to 2/3. That is,

$$\lim_{n\to\infty} \frac{a_{n+1}}{a_n} = \frac{2}{3}.$$

The Ratio Test gives convergence.

T/F 7 $\sum_{n=1}^{\infty} \sin \frac{1}{n^2}$ *converges.*

Solution. True. It is shown in nearly every introductory calculus text that

$$\lim_{x\to 0} \frac{\sin x}{x} = 1.$$

With that in mind, apply the Limit Comparison Test using the convergent series $\sum_{n=1}^{\infty} \frac{1}{n^2}$ to see that the original series converges also.

T/F 8 $\sum_{n=1}^{\infty} \cos \frac{1}{n^2}$ *converges.*

Solution. False. The terms approach $\cos 0 = 1 \neq 0$; apply the Divergence Test.

T/F 9 $\sum_{n=1}^{\infty} \frac{1}{n^{1+1/n}}$ *converges.*

Solution. False. See Gem 31. Note also that $\frac{1}{n^{1+1/n}} \geq \frac{1}{2}\frac{1}{n}$ and apply the Comparison Test.

T/F 10 *If p is a fixed real number, then $\sum_{n=1}^{\infty} 1/n^p$ converges if and only if $p > 1$.*

Solution. True. Theorem 1.59 (which was proven with the Integral Test).

T/F 11 *If $\sum_{n=1}^{\infty} a_n$ converges, then $\sum_{n=1}^{\infty} \frac{a_n}{n}$ converges.*

Solution. True. Apply Abel Test, Corollary 2.21.1, with a_n as given, $b_n = 1/n$.

T/F 12 *The series $\sum_{n=1}^{\infty} n^{\cos 3}$ converges.*

Solution. False. This is a p-series with $p = -\cos 3 < 1$.

T/F 13 $\sum_{n=1}^{\infty} \frac{n+1}{n+5}$ *converges to* $1/5$.

Solution. False. The terms approach 1, so the series in fact diverges.

T/F 14 $\sum_{n=2}^{\infty} 3(4)^{-n+1}$ *converges to* 1.

Solution. True. This is a geometric series with first term $3/4$ and ratio $1/4$. We use Theorem 1.39. The sum then is

$$\frac{3/4}{1 - 1/4} = 1.$$

T/F 15 *If $a_n > 0$ and the sequence of partial sums for $\sum_{n=1}^{\infty} a_n$ is bounded above, then the series converges.*

Solution. True. Theorem 1.25. A monotone increasing sequence that is bounded above must converge.

T/F 16 *If the sequence of partial sums for $\sum_{n=1}^{\infty} a_n$ is bounded, then the series converges.*

Solution. False. Consider the series $1 - 1 + 1 - 1 + 1 - 1 + \cdots$. Its partial sums oscillate between 1 and 0. Thus the partial sums are bounded but clearly do not converge.

T/F 17 *Every rearrangement of an absolutely convergent series has the same sum.*

Solution. True. Theorem 3.9.

T/F 18 *If the Strengthened Root Test indicates absolute convergence for a given series, then the Strengthened Ratio Test will also indicate absolute convergence for that series.*

Solution. False. Consider the series

$$\sum_{n=1}^{\infty} \frac{1}{k_n^n},$$

where k_n is 2 or 3 according as n is odd or even. The Strengthened Root Test clearly succeeds because the nth roots of terms are given by $1/k_n \leq 1/2 < 1$. However, the ratio of consecutive terms a_{n+1}/a_n does not yield to analysis because, for n even, we have a ratio of $3^n/2^{n+1} > 1$. In fact the ratios are not even bounded above, so the Strengthened Ratio Test fails.

T/F 19 *If* $\lim_{n\to\infty} a_n/b_n = \infty$ *and* $\sum_{n=1}^{\infty} b_n$ *diverges, then* $\sum_{n=1}^{\infty} a_n$ *diverges.*

Solution. False. Without assuming the terms are positive this does not follow from the Limit Comparison Test. Take $a_n = (-1)^n/\sqrt{n}$, $b_n = 1/(n\log n) + (-1)^n/n$. Then $\sum_{n=2}^{\infty} b_n$ diverges because it is the sum of a divergent positive series (by the Abel-Dini scale) and the convergent alternating harmonic series, but $\sum_{n=1}^{\infty} a_n$ is a convergent alternating series. Yet

$$\lim_{n\to\infty} \frac{a_n}{b_n} = \lim_{n\to\infty} \frac{(-1)^n/\sqrt{n}}{1/(n\log n) + (-1)^n/n} = \lim_{n\to\infty} \frac{\sqrt{n}}{1 + (-1)^n/\log n} = \infty.$$

T/F 20 *If* $\lim_{n\to\infty} a_n/b_n = 0$ *and* $\sum_{n=1}^{\infty} b_n$ *converges, then* $\sum_{n=1}^{\infty} a_n$ *converges.*

Solution. False. Interchanging the roles of a_n and b_n, this is the contrapositive of the previous item.

T/F 21 *If* $\sum_{n=1}^{\infty} a_n = A$ *and* $\sum_{n=1}^{\infty} |a_n| = B$ *and* A *and* B *are finite, then* $|A| = B$.

Solution. False. Take $a_n = (-1/2)^n$. Then using Theorem 1.39,

$$\sum_{n=1}^{\infty}(-1/2)^n = -\frac{1}{3}; \ \sum_{n=1}^{\infty}(1/2)^n = 1.$$

T/F 22 *If* $\sum_{n=1}^{\infty} a_n$ *diverges and* $a_n > 0$, *then* $\sum_{n=1}^{\infty} \frac{a_n}{1+a_n}$ *diverges also.*

Solution. True. In the case that $a_n \to 0$, we eventually have

$$\frac{a_n}{1+a_n} > \frac{a_n}{2},$$

and the Comparison Test applies. If on the other hand $a_n \not\to 0$, then the terms $\frac{a_n}{1+a_n}$ do not go to 0 either and the Divergence Test applies.

T/F 23 *If both $\sum_{n=1}^{\infty} a_n$ and $\sum_{n=1}^{\infty} b_n$ diverge, then $\sum_{n=1}^{\infty}(a_n + b_n)$ diverges.*

Solution. False. Take $a_n = 1, b_n = -1$.

T/F 24 *If both $\sum_{n=1}^{\infty} a_n$ and $\sum_{n=1}^{\infty} b_n$ diverge, then $\sum_{n=1}^{\infty}(a_n + b_n)$ converges.*

Solution. False. Take $a_n = b_n = 1$.

T/F 25 *If $\sum_{n=1}^{\infty} a_n$ converges and $\sum_{n=1}^{\infty} b_n$ diverges, then $\sum_{n=1}^{\infty}(a_n + b_n)$ diverges.*

Solution. True. Let A_n, B_n, C_n be the partial sums of the three series, respectively, and assume to the contrary that $\{C_n\}$ did converge. Then, since $B_n = C_n - A_n$ and the latter two sequences converge, $\{B_n\}$ must converge also, contrary to assumption.

T/F 26 *If $\sum_{n=1}^{\infty} a_n$ converges and $\sum_{n=1}^{\infty} b_n$ diverges, then $\sum_{n=1}^{\infty}(a_n + b_n)$ converges.*

Solution. False. This would contradict the previous item (T/F 25).

T/F 27 *If $a_n > 0$ and $\sum_{n=1}^{\infty} a_n$ converges, then $\sum_{n=1}^{\infty} 1/a_n$ diverges.*

Solution. True. Using Theorem 1.24, $\sum_{n=1}^{\infty} a_n$ converges implies that $a_n \to 0$, hence $1/a_n \to \infty$. Then $\sum_{n=1}^{\infty} 1/a_n$ diverges by the Divergence Test.

T/F 28 *If $a_n > 0$ and $\sum_{n=1}^{\infty} a_n$ diverges, then $\sum_{n=1}^{\infty} 1/a_n$ converges.*

Solution. False. Take $a_n = 1$.

T/F 29 *If $\lim_{n\to\infty} a_n/b_n = 1$ and $\sum_{n=1}^{\infty} b_n$ converges, then $\sum_{n=1}^{\infty} a_n$ converges.*

Solution. False. Again, the Limit Comparison Test does not apply because the terms need not be nonnegative. Take $b_n = (-1)^n/\sqrt{n}$, the terms of a convergent alternating series, and take $a_n = b_n + 1/n$. Then $\sum_{n=1}^{\infty} a_n$ is divergent, because it is formed as the sum of a convergent series and a divergent (harmonic) series. We now check

$$\begin{aligned}\lim_{n\to\infty} \frac{a_n}{b_n} &= \lim_{n\to\infty} \frac{(-1)^n/\sqrt{n} + 1/n}{(-1)^n/\sqrt{n}} \\ &= \lim_{n\to\infty} \left(1 + \frac{(-1)^n}{\sqrt{n}}\right) \\ &= 1.\end{aligned}$$

T/F 30 *If* $\lim_{n\to\infty} a_n = 0$ *and* $\sum_{n=1}^{\infty} b_n$ *converges, then* $\sum_{n=1}^{\infty} a_n b_n$ *converges.*

Solution. False. Take $a_n = b_n = (-1)^n/\sqrt{n}$.

T/F 31 *If* $\lim_{n\to\infty} a_n = 0$, $0 < a_{n+1} \le a_n$ *and* $\sum_{n=1}^{\infty} b_n$ *converges, then* $\sum_{n=1}^{\infty} a_n b_n$ *converges.*

Solution. True. Apply Abel's Test.

T/F 32 *If* $\lim_{n\to\infty} a_n = 0$ *and* $\sum_{n=1}^{\infty} b_n$ *converges absolutely, then* $\sum_{n=1}^{\infty} a_n b_n$ *converges.*

Solution. True. Apply the Comparison Test to $\sum_{n=1}^{\infty} |b_n|$, $\sum_{n=1}^{\infty} |a_n b_n|$, noticing that eventually we must have $|a_n| < 1$.

T/F 33 *If* $a_n \ge 0$, *then* $\sum_{n=1}^{\infty} a_n$ *does not converge conditionally.*

Solution. True. Since $\sum_{n=1}^{\infty} |a_n|$, $\sum_{n=1}^{\infty} a_n$ are in fact the same series, they cannot have different end behavior.

T/F 34 *The Alternating Series Test is a test for conditional convergence.*

Solution. False. While the Alternating Series Test is capable of showing that conditionally convergent series are convergent, it does not check that the associated series with all terms positive is divergent. Thus it cannot distinguish absolutely convergent alternating series from conditionally convergent alternating series.

T/F 35 *If* $\sum_{n=1}^{\infty} a_n b_n$ *converges, then* $\sum_{n=1}^{\infty} a_n$ *and* $\sum_{n=1}^{\infty} b_n$ *converge.*

Solution. False. Take $a_n = b_n = 1/n$.

T/F 36 *If* $a_n > 0$ *and* $\sum_{n=1}^{\infty} a_n$ *converges, then* $\sum_{n=1}^{\infty} \sin a_n$ *converges.*

Solution. True. It is shown in nearly every introductory calculus text that

$$\lim_{x\to 0} \frac{\sin x}{x} = 1.$$

With that in mind, apply the Limit Comparison Test, noting that $a_n \to 0$.

T/F 37 *If* $\sum_{n=1}^{\infty} a_n$ *converges, then* $\sum_{n=1}^{\infty} \cos a_n$ *converges.*

Solution. False. By Theorem 1.44, $a_n \to 0$. Thus $\cos a_n \to \cos 0 = 1 \ne 0$, and the Divergence Test applies.

T/F 38 *If* $0 < a_{n+1} < a_n$ *for all* n *and* $\lim_{n\to\infty} a_n = 0$, *then*

$$\sum_{n=1}^{\infty}(-1)^{n+1}a_n$$

converges and has sum S *satisfying* $|S_n - S| < a_{n+1}$.

Solution. True. This is the Alternating Series Test (Theorem 1.75).

T/F 39

$$\sum_{n=1}^{\infty}\left(1-\frac{1}{n}\right)^n$$

is a divergent series.

Solution. True. The terms do not tend to 0. (In fact they tend to $1/e$.)

T/F 40 *The Root Test cannot alone be used to determine conditional convergence.*

Solution. True. In order to verify conditional convergence one must show 1) convergence for the series as given, and 2) that the series does not converge absolutely. The Root test may only give information as regards the second requirement.

T/F 41 *The Ratio Test cannot alone be used to determine conditional convergence.*

Solution. True. In order to verify conditional convergence one must show 1) convergence for the series as given, and 2) that the series does not converge absolutely. The Ratio test may only give information as regards the second requirement.

T/F 42

$$\sum_{n=1}^{\infty}\frac{1}{n^n} \geq 2.$$

Solution. False. We simply observe that

$$\sum_{n=1}^{\infty}\frac{1}{n^n} = 1 + \sum_{n=2}^{\infty}\frac{1}{n^n} < 1 + \sum_{n=2}^{\infty}\frac{1}{2^n} = 3/2.$$

T/F 43 *The Ratio Test cannot be used to determine the convergence or divergence of the series*

$$\sum_{n=1}^{\infty}\frac{n}{n^3+5}.$$

Solution. True. The limiting ratio is 1. In fact the same would happen with any series with terms algebraic in n.

T/F 44 *If $\sum_{n=1}^{\infty} a_n$ converges, then $\sum_{n=1}^{\infty}(-1)^n a_n$ converges also.*

Solution. False. Consider $a_n = (-1)^n/n$.

T/F 45

$$\frac{1}{2} + \left(\frac{1}{2}\right)^2 + \left(\frac{1}{2}\right)^3 + \cdots + \left(\frac{1}{2}\right)^{10000} < 1.$$

Solution. True.

$$\sum_{n=1}^{10000} (1/2)^n < \sum_{n=1}^{\infty} (1/2)^n = 1,$$

using Theorem 1.39.

T/F 46 *If $\sum_{n=1}^{\infty} b_n$ converges and if $0 \le a_{n+k} \le b_n$ for some $k \ge 1000$ and every $n \ge 1000$, then $\sum_{n=1}^{\infty}(-1)^n a_n$ converges absolutely.*

Solution. True. Since the behavior of a series is a property of any of its tails, we clearly have $\sum_{n=1}^{\infty} b_{n+k}$ to be convergent since $\sum_{n=1}^{\infty} b_n$ is. The Comparison Test applies.

T/F 47

$$\sum_{n=1}^{\infty} \frac{1}{\log(n^6 + 1)}$$

converges.

Solution. False. Apply the Limit Comparison Test against

$$\sum_{n=1}^{\infty} \frac{1}{\log(n^6)} = \sum_{n=1}^{\infty} \frac{1}{6\log(n)}$$

which in turn diverges by comparison with the harmonic series.

T/F 48 *If $a_n \le b_n \le 0$ and $\sum_{n=1}^{\infty} b_n$ diverges, then $\sum_{n=1}^{\infty} a_n$ diverges also.*

Solution. True. Changing the sign of every term in a series does not affect its convergence. We apply the Comparison Test to the series $\sum_{n=1}^{\infty} -b_n$ and $\sum_{n=1}^{\infty} -a_n$.

T/F 49 *If $a_n \leq b_n \leq 0$ and $\sum_{n=1}^{\infty} a_n$ converges, then $\sum_{n=1}^{\infty} b_n$ converges also.*

Solution. True. Changing the sign of every term in a series does not affect its convergence. We apply the Comparison Test to the series $\sum_{n=1}^{\infty} -b_n$ and $\sum_{n=1}^{\infty} -a_n$.

T/F 50 *If $a_n > 0$ and $\sum_{n=1}^{\infty} a_n$ converges, then $\sum_{n=1}^{\infty}(-1)^n a_n$ converges also.*

Solution. True. By assumption it converges absolutely, hence it is also convergent.

T/F 51 *If the condition $0 \leq a_{n+1} \leq a_n$ fails (even eventually) when applying the Alternating Series Test, then the series $\sum_{n=1}^{\infty}(-1)^n a_n$ diverges.*

Solution. False. Consider $a_n = \frac{1}{k_n^n}$, where k_n is 2 or 3 according as n is odd or even. By the Strengthened Root Test, $\sum_{n=1}^{\infty}(-1)^n a_n$ converges absolutely. However, $a_{n+1} \leq a_n$ never holds for even n.

T/F 52 *If the condition $\lim_{n\to\infty} a_n = 0$ fails when applying the Alternating Series Test, then the series $\sum_{n=1}^{\infty}(-1)^n a_n$ diverges.*

Solution. True. Use the Divergence Test.

T/F 53 *If the Ratio Test indicates absolute convergence for a particular series, then the Root Test will also indicate absolute convergence for that series.*

Solution. True. Suppose that the Ratio Test will show convergence for $\sum_{n=1}^{\infty} a_n$, where without loss of generality $a_n \geq 0$. Then consider $r_n = a_n/a_{n-1}$. Since the Ratio Test is decisive, there exists a constant $c < 1$ such that eventually $r_n < c$. Let N be large enough that $r_n < c$ holds for $n \geq N$. Then, for $n \geq N$,

$$a_n^{1/n} = (a_N(a_n/a_N))^{1/n} < (a_N c^{n-N})^{1/n} = c\left(\frac{a_N}{c^N}\right)^{1/n}.$$

Now the expression in the parentheses is independent of n, so that as n grows the exponential expression will approach 1. Thus, the $a_n^{1/n} < c < 1$ eventually, and the Root Test is decisive.

T/F 54

$$\sum_{n=1}^{\infty} \frac{n+2}{n+4} = \infty.$$

Solution. True. Since the series is positive, it will suffice to show that the series does not converge. This follows from the Divergence Test because the terms do not approach 0.

T/F 55 *If* $\sum_{n=1}^{\infty} a_n = 7$ *and* $a_1 = a_2 = 3$, *then* $\sum_{n=3}^{\infty} a_n = 1$.

Solution. True. Let S_n be the nth partial sum of the first series and let T_n ($T_1 = 0,\ T_2 = 0$), be the nth partial sum of the second series. For $n \geq 3$, $S_n = T_n + 6$. Since $S_n \to 7$, $T_n \to 1$.

T/F 56 $3/2, 4/3, 5/4, 6/5, \ldots, (n+1)/n, \ldots$ *is a convergent series.*

Solution. False. It is a convergent sequence. In everyday language sequence and series are used almost interchangebly, but not in mathematics. In addition we note that the series $\sum_{n=2}^{\infty} \frac{n+1}{n}$ diverges since $a_n = \frac{n+1}{n} \to 1$.

T/F 57 *The first three terms in the sequence of partial sums associated with* $\sum_{n=1}^{\infty}(1/2)^n$ *are 1/2,3/4,7/8.*

Solution. True. This is a direct computation.

T/F 58

$$\sum_{n=2}^{\infty}\left(\frac{1}{3n+1} - \frac{1}{3n+4}\right) = \frac{1}{7}.$$

Solution. True. Observe that the series telescopes. The nth partial sum is $1/7 - 1/(3n+4)$, which tends to 1/7.

T/F 59 *If* $\lim_{n\to\infty} a_{2n} = 3$ *and* $\lim_{n\to\infty} a_{2n+1} = 3$, *then* $\lim_{n\to\infty} a_n = 3$.

Solution. True. Fix an $\epsilon > 0$. Then we can find an N_1, N_2 such that for $n > N_1$, $|a_{2n} - 3| < \epsilon$ and for $n > N_2$, $|a_{2n+1} - 3| < \epsilon$. Then $|a_n - 3| < \epsilon$ for all $n > \max\{2N_1, 2N_2 + 1\}$.

T/F 60

$$\lim_{n\to\infty} a_n = 0 \Leftrightarrow \lim_{n\to\infty} |a_n| = 0.$$

Solution. True. We simply notice that both statements are, by definition, equivalent to the statement that for each $\epsilon > 0$ we have $|a_n| < \epsilon$ eventually.

T/F 61 *If* $0 \leq a_n \leq b_n$ *and* $\sum_{n=1}^{\infty} b_n$ *diverges, then* $\sum_{n=1}^{\infty} a_n$ *diverges also.*

Solution. False. Take $a_n = 1/n^2$, $b_n = 1$.

T/F 62 *If a series converges, then no subsequence of the sequence of partial sums can be unbounded.*

Solution. True. Since the series converges, its sequence of partial sums converges. Thus any subsequence would also converge to the same value.

T/F 63 *The integral test is of no value in testing* $\sum_{n=1}^{\infty} \sin n/n$ *for convergence or divergence.*

Solution. True. The Integral Test applies only to positive series, and the series in question changes signs erratically.

T/F 64 *The sum of the infinite series* $\sum_{n=1}^{\infty} a_n$ *is the limit*

$$S = \lim_{k\to\infty} \sum_{n=1}^{k} a_n,$$

whenever this limit exists.

Solution. True. See Definitions 1.34 and 1.35.

T/F 65 $\sum_{n=1}^{\infty} a_n$ *converges if and only if* $\sum_{n=p}^{\infty} a_n$ *converges for every* p.

Solution. True. Each tail of a series will have the same limiting behavior.

T/F 66 *The Ratio Test is of no value in testing* $\sum_{n=1}^{\infty} 1/\sqrt{n(n+1)(n+2)}$ *for convergence or divergence.*

Solution. True. Again, a series with algebraic terms cannot be decided with the Ratio Test. We note that the ratios are given by

$$\frac{1/\sqrt{(n+1)(n+2)(n+3)}}{1/\sqrt{n(n+1)(n+2)}} = \frac{\sqrt{n}}{\sqrt{n+3}} \to 1.$$

T/F 67

$$\sum_{n=2}^{\infty} \left(1 - \frac{1}{n^2}\right)^{n^2}$$

diverges.

Solution. True. The terms do not tend to 0, so the Divergence Test applies. (In fact they tend to $1/e$.)

T/F 68 *If* $a_n > 0$ *for all* n *and if* $\sum_{n=1}^{\infty} a_n$ *converges, then*

$$\lim_{n\to\infty} \frac{a_{n+1}}{a_n} = L < 1.$$

Solution. False. Consider any convergent p-series. (It is also possible for the limit not to exist.)

T/F 69 *If for some constant c we have $ca_n \geq 1/n$, then necessarily $\sum_{n=1}^{\infty} a_n$ diverges.*

Solution. True. Apply the Comparison Test using the harmonic series.

T/F 70 *The series $\sum_{n=2}^{\infty}(-1)^n$ has as its sequence of partial sums $1, 0, 1, 0, \ldots$.*

Solution. True. This is a direct computation.

T/F 71 *If $\lim_{n\to\infty} a_n = 0$, then*

$$\sum_{n=1}^{\infty} \frac{a_n + 2}{a_n^2 + 1} = 2.$$

Solution. False. The terms approach $2 \neq 0$, so that the series actually diverges by the Divergence Test.

T/F 72 *If $\lim_{n\to\infty} a_n = 0$, then*

$$\sum_{n=1}^{\infty} \frac{a_n}{a_n^2 + n^2}$$

converges absolutely.

Solution. True. Since $a_n \to 0$ we must eventually have $|a_n| < 1$. Then

$$\left|\frac{a_n}{a_n^2 + n^2}\right| = \frac{|a_n|}{a_n^2 + n^2} \leq \frac{1}{a_n^2 + n^2} \leq \frac{1}{n^2}.$$

We then apply the Comparison Test.

T/F 73

$$\sum_{n=1}^{\infty} n\left(\frac{1}{2}\right)^n = 2.$$

Solution. True. We know that

$$\frac{1}{1-x} = \sum_{n=0}^{\infty} x^n.$$

Taking a derivative,

$$\frac{1}{(1-x)^2} = \sum_{n=1}^{\infty} nx^{n-1}$$

and then multiplying by x, we get

$$\frac{x}{(1-x)^2} = \sum_{n=1}^{\infty} nx^n.$$

The result then follows from taking $x = 1/2$.

T/F 74

$$\sum_{k=1}^{\infty} \left(Arctan\frac{1}{k} - Arctan\frac{1}{k+1} \right) = \frac{\pi}{4}.$$

Solution. True. Notice that the series telescopes.

T/F 75 *The Ratio Test is of no help in determining whether*

$$\sum_{n=1}^{\infty} \frac{(2n)!}{(n!)^2}.$$

converges or diverges.

Solution. False. The ratio of consecutive terms is

$$\frac{(2n+2)!/((n+1)!)^2}{(2n)!/(n!)^2} = \frac{(2n+2)(2n+1)}{(n+1)^2} \to 4 > 1.$$

The series diverges by the Ratio Test.

T/F 76 *If* $|a_{n+1}/a_n| > 1$ *for all n, then* $\sum_{n=1}^{\infty} |a_n|$ *diverges.*

Solution. True. We see that the terms $|a_n|$ are actually increasing, and thus are not tending to 0.

T/F 77 *If* $|a_{n+1}/a_n| < 1$ *for all n, then* $\sum_{n=1}^{\infty} |a_n|$ *converges.*

Solution. False. Consider the harmonic series.

T/F 78 *If* $s_n = 3 + 2^{-n}$ *then* $\lim_{n\to\infty} s_n = 4$.

Solution. False. The limit is 3.

T/F 79 *The first three terms in the sequence of partial sums associated to the series* $\sum_{n=1}^{\infty}(-1/2)^n$ *are* $-1/2, 1/4, -1/8$.

Solution. False. These are the first three terms of the series. The first three partial sums are $-1/2, -1/4, -3/8$.

T/F 80

$$\sum_{n=1}^{\infty} 3^{-n}2^{n+1} = 4.$$

Solution. True. This is a geometric series with first term 4/3 and common ratio 2/3. We apply Theorem 1.39.

T/F 81 *Every partial sum of*

$$\sum_{n=1}^{\infty} \left(\frac{1}{n(n+1)} \right)$$

is less than 1.

Solution. True. Writing

$$\frac{1}{n(n+1)} = \frac{1}{n} - \frac{1}{n+1}$$

we see that the series telescopes, so that the partial sums have the form $1 - 1/(n+1) < 1$.

T/F 82 *The Integral Test can be used to show that*

$$\sum_{n=3}^{\infty} \frac{(-1)^n}{n \log n \log \log n}$$

does not converge absolutely.

Solution. True. The series

$$\sum_{n=3}^{\infty} \frac{1}{n \log n \log \log n}$$

diverges by the Abel-Dini scale, which was proven by the Integral Test.

T/F 83 *The insertion or removal of a finite number of terms in a series cannot affect its convergence or divergence, although it may affect its sum.*

Solution. True. Regardless of what finite number of terms are changed, some tail sufficiently far out is unaffected, so that the convergence or divergence is preserved.

T/F 84 *If a series converges, then any summation of a subset of the terms will form a convergent series.*

Solution. False. Let $\sum_{n=1}^{\infty} a_n$ by any divergent positive series whose terms tend to zero. (The harmonic series is an example.) Then consider the series $a_1 - a_1 + a_2 - a_2 + a_3 - a_3 + \cdots$. This series converges to 0, but the sum of only the positive terms diverges properly.

As an explicit counterexample, the alternating harmonic series converges but the subseries formed by the positive terms diverges.

T/F 85 *If $\sum_{n=1}^{\infty} a_n$ converges and $\{b_n\}_{n=1}^{\infty}$ is a bounded sequence, then $\sum_{n=1}^{\infty} a_n b_n$ converges.*

Solution. False. Let $a_n = b_n = \frac{(-1)^n}{\sqrt{n}}$.

T/F 86 *If $\sum_{n=1}^{\infty} a_n$ diverges, then $\sum_{n=1}^{\infty} |a_n|$ must diverge.*

Solution. True. An absolutely convergent series must converge by Theorem 1.70. This states the contrapositive.

T/F 87 *If $a_n = 3/n^2$, then $\sum_{n=1}^{\infty}(a_n - a_{n+1})$ converges to zero.*

Solution. False. The series telescopes, and it is easy to check that the sum is 3.

T/F 88 *A series representation for the decimal* 0.999999... *can be given by*

$$\sum_{n=1}^{\infty} 9\left(\frac{1}{10}\right)^n$$

where the sum is exactly 1.0, *so* $0.999999999999\ldots = 1$.

Solution. True. The series can be expanded as $0.9 + 0.09 + 0.009 + 0.0009 + \cdots$. As for the sum, we apply Theorem 1.39.

T/F 89 *If $\lim_{n\to\infty} a_n$ does not exist, then we can always correctly claim that $\sum_{n=1}^{\infty} a_n$ diverges.*

Solution. True. Apply the Divergence Test.

T/F 90 *A convergent series is either absolutely convergent or conditionally convergent.*

Solution. True. Either $\sum_{n=0}^{\infty} |a_n|$ converges, in which case we have absolute convergence, or else it doesn't, in which case we have conditional convergence.

T/F 91 *The alternating harmonic series cannot be rearranged to diverge to* $-\infty$.

Solution. False. See Theorem 3.7.

T/F 92 *If* $a_n > 0$ *and* $\sqrt[n]{a_n} < 1$ *for all* n*, then* $\sum_{n=1}^{\infty} a_n$ *converges.*

Solution. False. Take $a_n = (1 - 1/n)^n$.

T/F 93 *If* $a_n > 0$ *and* $\sqrt[n]{a_n} \geq 1$ *for all* n*, then* $\sum_{n=1}^{\infty} a_n$ *diverges.*

Solution. True. We would then have $a_n \geq 1$, so that the Divergence Test applies.

T/F 94 *If* $a_n > 0$ *and* $\sqrt[n]{a_n} < r < 1$ *for all* n *and some constant* r*, then* $\sum_{n=1}^{\infty} a_n$ *converges.*

Solution. True. This follows from the Root Test.

T/F 95 *If* $\lim_{n\to\infty} |a_{n+1}/a_n| = L < 1$ *for some constant* L*, then* $\lim_{n\to\infty} a_n = 0$

Solution. True. We first apply the Ratio Test to see that $\sum_{n=1}^{\infty} a_n$ converges and then apply Theorem 1.24.

T/F 96 *If* $\sum_{n=1}^{\infty} a_n^2$ *converges, then* $\sum_{n=1}^{\infty} \frac{a_n}{n}$ *converges.*

Solution. True. See Gem 1.

T/F 97 *If* $a_n \geq 0$ *and* $\sum_{n=1}^{\infty} a_n$ *converges, then* $\sum_{n=1}^{\infty} \frac{\sqrt{a_n}}{n^p}$ *converges for* $p > 1/2$.

Solution. True. See Gem 2 and note that $\sum_{n=1}^{\infty} 1/n^{2p}$ converges for $p > 1/2$.

T/F 98 *If* $a_n \geq 0$ *and* $\sum_{n=1}^{\infty} a_n$ *diverges, then* $\sum_{n=1}^{\infty} \frac{a_n}{1+na_n}$ *diverges.*

Solution. False. Consider $a_n = \sqrt{n}$ in the case that $\sqrt{n}$ is an integer and $a_n = 0$ otherwise.

T/F 99 *If* $a_n \geq 0$ *and* $\sum_{n=1}^{\infty} a_n$ *diverges, then* $\sum_{n=1}^{\infty} \frac{a_n}{1+na_n}$ *converges.*

Solution. False. Consider $a_n = 1$.

T/F 100 *If* $a_n \geq 0$ *and* $\sum_{n=1}^{\infty} a_n$ *diverges, then* $\sum_{n=1}^{\infty} \frac{a_n}{1+a_n^2}$ *converges.*

Solution. *False.* Consider $a_n = n$.

T/F 101 *If* $a_n \geq 0$ *and* $\sum_{n=1}^{\infty} a_n$ *diverges, then* $\sum_{n=1}^{\infty} \frac{a_n}{1+a_n^2}$ *diverges.*

Solution. *False.* Consider $a_n = 2^n$.

B

What's Harmonic about the Harmonic Series?*

Why is $1+\frac{1}{2}+\frac{1}{3}+\frac{1}{4}+\cdots$ called the harmonic series? A simple answer would be that each term of the series, after the first, is the harmonic mean of its two nearest neighbors. This response, however, is likely to raise more questions than it answers: What is the harmonic mean? Where and when did it originate? Is there any connection to musical harmony?

B.1 The Harmonic Mean

The arithmetic mean of a and b, $A=\frac{a+b}{2}$, should be familiar to all mathematics students, and the geometric mean, $G=\sqrt{ab}$, is almost as well known. The harmonic mean, however, has been called "the neglected mean" [4], [8] because it appears so infrequently in core mathematics courses. The *harmonic mean* of two numbers a and b is defined by the formula $H=\frac{2ab}{a+b}$ or, equivalently, $\frac{2}{H}=\frac{1}{a}+\frac{1}{b}$. It follows that, for $n>1$, $\frac{1}{n}$ is the harmonic mean of $\frac{1}{n-1}$ and $\frac{1}{n+1}$. More generally, if x, $x+y$, and $x+2y$ are in arithmetic progression, then their reciprocals, $a=\frac{1}{x}$, $H=\frac{1}{x+y}$, and $b=\frac{1}{x+2y}$ are said to be in *harmonic progression.*

The origins of the harmonic mean can be traced back to the ancient Pythagoreans. According to the commentator Proclus (410–485), Pythagoras himself (6th century BC) learned about three means—arithmetic, geometric, and subcontrary (later called harmonic)—while visiting Mesopotamia [1]. According to this ancient tradition, Pythagoras also knew a "golden proportion" relating these means. Given two numbers, the first is to their arithmetic mean as their harmonic mean is to the second. That is, $\frac{a}{(a+b)/2}=\frac{2ab/(a+b)}{b}$. Another way to express this is that the harmonic mean is the ratio of the square of the geometric mean to the arithmetic mean ($H=G^2/A$).

"Arithmetic" in the time of the ancient Greeks consisted of the study of properties of whole numbers and rational numbers. The arithmetic mean of two rational numbers must

*Reprinted with permission from The College Mathathematics Journal.

also be rational but, for most pairs of integers, the geometric mean is irrational, and the Greeks relegated such magnitudes to the realm of geometry. These facts may help to explain the names of two of our three means. Later, we'll examine a possible explanation for the name of the harmonic mean.

In passing, we note that the term "harmonic range of points" is also derived from the harmonic mean. Collinear points A, B, C, D are said to form a *harmonic range* if and only if their cross ratio, $(AB, CD) = (\overline{AC}/\overline{CB})/(\overline{AD}/\overline{DB})$, is -1. It can be shown [5] that this is equivalent to $2/\overline{AB} = 1/\overline{AC} + 1/\overline{AD}$. That is, the length of AB is the harmonic mean of the lengths of AC and AD.

B.2 A Musical Connection

Pythagoras (or one of his disciples) is also credited with giving Western Europe its first theory of music, based on musical intervals that can be expressed in terms of numerical ratios of small whole numbers. Among these are the *octave* (2:1), the *fifth* (3:2), and the *fourth* (4:3). Here the numbers 1, 2, 3, and 4 are proportional to the frequencies of the tones, a larger number corresponding to a higher pitch. The tone corresponding to 1 is called the *tonic*. The names fourth, fifth, and octave come from the ordering of tones on an 8-tone diatonic scale.

The ratio 2:1 for an octave may also be expressed as 12:6, so that the tonic has relative frequency 6. Suppose that the ancient Pythagoreans wanted to introduce a new tone, roughly midway between the tonic and the octave. The geometric mean of 6 and 12 is irrational, so a musical tone based on it would have been unacceptable. The arithmetic mean of 6 and 12 is 9, leading to the ratios 9:6 = 3:2 and 12:9 = 4:3. In such a scale the new tone (corresponding to a relative frequency of 9) would be a fifth above the tonic and the octave would be a fourth above the new tone.

On the other hand, the harmonic mean of 6 and 12 is 8, producing ratios 8:6 = 4:3 and 12:8 = 3:2. Now the new tone is a fourth above the tonic, with the octave a fifth above that. A surviving fragment of the work of Archytas of Tarentum (ca. 350 BC) states, "There are three means in music: one is the arithmetic, the second is the geometric, and the third is the subcontrary, which they call harmonic" [7]. The term subcontrary may refer to the fact that a tone based on this mean reverses the order of the two fundamental musical intervals in a scale.

It is believed that Archytas or one of his contemporaries gave the name "harmonic" to the subcontrary mean because it produced a division of the octave in which the middle tone stood in the most harmonious relationship to the tonic and the octave. Even today, many musicians prefer a scale in which a fourth is followed by a fifth. (A more complete discussion of Greek musical theory may be found in [2] and [9].)

B.3 Divergence of the Harmonic Series

The earliest recorded proof that the harmonic series diverges is found in a treatise by the 14th century Parisian scholar, Nicole Oresme. In the third proposition of his work, *Ques-*

tions on the geometry of Euclid (ca. 1350) Oresme asserted that, "It is possible that an addition could be made, though not proportionally, to any quantity by ratios of lesser inequality, and yet the whole would become infinite" [6]. His example consists of adding to a one-foot quantity one-half of a foot, then one-third of a foot, then one-fourth and one-fifth "and so on to infinity." His proof is essentially the one that we often show our students today. He notes that $1/4$ plus $1/3$ is greater than $1/2$; the sum of the terms from $1/5$ to $1/8$ is greater than $1/2$; and so on, obtaining infinitely many parts "of which any one will be greater than one-half foot and the whole will be infinite."

Oresme's result was rediscovered by Pietro Mengoli in 1672 and Jacques Bernoulli in 1689 [1]. The proof given by Bernoulli in *Ars conjectandi* has been characterized by William Dunham [3] as "entirely different, yet equally ingenious." According to the *Oxford English Dictionary*, the name "harmonic series" first appeared in *Chambers Cyclopedia* (1727–51): "Harmonical series is a series of many numbers in continual harmonical proportion."

The roots of the harmonic series can be traced back to some of the earliest mathematical traditions in Western culture. Based on the harmonic mean, it is related to harmonics in both music and geometry. This concept deserves its place as a fundamental idea of mathematical thought.

References

1. Carl B. Boyer, *A History of Mathematics*, Wiley, 1968, pp. 61, 406.
2. Richard L. Crocker, Pythagorean mathematics and music, *Journal of Aesthetics and Art Criticism*, 22 (1963–64), 189–198 and 325–335.
3. William Dunham, The Bernoullis and the harmonic series, *College Mathematics Journal*, 18 (1987), 18–23.
4. Joseph L. Ercolano, Remarks on the neglected mean, *Mathematics Teacher*, 67 (1973), 253–255.
5. Howard Eves, *A Survey of Geometry*, Revised Ed., Allyn & Bacon, 1972, p. 83.
6. Edward Grant, ed., *A Source Book in Medieval Science*, Harvard University Press, 1974, p. 135.
7. Thomas L. Heath, *A History of Greek Mathematics* (vol. 1), Dover reprint 1981, p. 85.
8. Alfred S. Posamentier, The neglected mean, *School Science and Mathematics*, 77 (1977), 339–344.
9. B. L. van der Waerden, *Science Awakening*, Wiley, 1963, pp. 149–159.

C

References

Despite the best efforts from the authors of this book, we have done little more than offer a spoonful of information from the mountain that is infinite series. To the extent that our goal is to present the generally interesting and beautiful results of infinite series, no single book could be sufficiently inclusive. The world of infinite series, even series of real constants, is just too big. We first give a selection of books on the subject. Some of these books are readily available and some are quite hard to locate. We also give a short list of textbooks on calculus, or advanced calculus, and recommend them as providing the basics, or more, on infinite sequences and series. Nearly all calculus books, beginning or advanced, as well as real analysis books present the basic information on sequences and series hence any such book you select should prove helpful.

Students of mathematics can learn valuable principles by way of problem solving. It is in this spirit that we provide a list of problem books each of which contains problems on infinite series. The book by Kaczor and Nowak contains an especially nice collection of infinite series problems, with solutions, and they vary from the relatively simple to quite challenging. The reader who is interested in discovering more of the gems within the study of infinite series may wish to read some of the articles, as found in three journals published by the MAA. In virtually all cases, the material in this book provides all prerequisite knowledge for these articles, so by all means visit your local university library and find out more. As mentioned in the preface, you may also use JSTOR (Journal Storage) to access our list of journal articles. The URL for JSTOR is http://www.jstor.org/.

Books on Infinite Series

- Bromwich, T. J. I'A., *An Introduction to the Theory of Infinite Series*, Macmillan and Co. Ltd., London, 1931

  - This textbook is outstanding in its thoroughness, giving adequate attention to multiple series, infinite products, function series, power and trigonometric series, complex series, and nonconvergent series. However, the style and length make this text more suitable as a reference for specific topics than for cover-to-cover study.

- Davis, Harold T., *The Summation of Series*, Principia Press of Trinity University (TX), 1962

  - This author does an excellent job of collecting in one place numerous techniques for evaluating, summing, and estimating infinite series and other related limit objects. The intersection of this book with ours occurs in Chapter 5, "Infinite Sums." Even if Chapters 4 and 6, dealing directly with reference tables and their effective use, are of limited relevance today, Chapters 1 ("The Calculus of Finite Differences"), 2 ("The Gamma and Psi Functions"), and 3 ("Other Methods of Summation") stand out.

- Fichtenholz, G. M., *Infinite Series*, Gordon and Breach Science Publishers, Inc. New York

  - The original work of Fichtenholz is available from Gordon and Breach in English translation. This translation is the work of Richard Silverman, who has also adapted and expanded upon them extensively, and is available in two volumes: *Rudiments* and *Ramifications*.

- Fort, Tomlinson, *Infinite Series*, Oxford University Press, 1930

- Hardy, G. H., *Divergent Series*, American Mathematical Society, Providence, 1991

  - This book is of a different nature than the others in this collection, as you might imagine from its title. We made mention earlier of a variety of "alternate" definitions of series convergence and, in the case of convergence, definitions of what is meant by the sum. This book is a remarkably thorough exploration of many of the important generalizations and modifications of the definitions and treatment presented here. The text includes, for example, Nörlund means, Euler means, Abelian means, Riesz means, Hölder means, and Cesàro summation. Certainly an advanced text, but a good reference.

- Hirschman, Jr., Isidore Isaac, *Infinite Series*, Holt, Rinehart and Winston (NY), 1962

  - This book is friendly and efficient in its treatment, and the exercises and problems are plentiful and very appropriate in terms of difficulty. The reader interested in learning about series of functions, which we have only alluded to, would find the transition from here to Hirschman smooth and rewarding.

- Hyslop, James Morton, *Infinite Series, 5th ed.*, Interscience Publisher, Inc. (NY), 1959

- Jolley, Leonard Benjamin William, *Summation of Series*, Dover (NY), 2004

- This a book-length table of interesting summations, including finite and infinite examples of constants and functions. The examples are sorted by type to facilitate looking up a particular series to see what is known about it; if in the course of your work you run into a summation expression, chances are it is in Jolley.

- Knopp, Konrad, *Infinite Sequences and Series*, Dover, (NY), 1956
- Knopp, Konrad, *Theory and Application of Infinite Series*, Dover, (NY), 1990
  - The books of Konrad Knopp have been considered classic introductions to infinite series for decades. The treatment is straightforward and extensive, as well as being quite readable.
- Markushevich, A. I., *Infinite Series*, D. C. Heath & Company, Boston (MA), 1967
  - This is a short book that focuses less on theory than most of the books in this bibliography, certainly less than we have, and more on classical numerical methods for estimating the sum of convergent series with high accuracy. Markushevich focuses his attention on series representations for functions, especially the power series. Even though many of the applications specifically discussed have been rendered obsolete by calculators (the book was written during the time of logarithm tables), the estimation techniques discussed are still valuable.
- Rainville, Earl D., *Infinite Series*, The Macmillan Company (NY), 1967
  - This nice text goes quickly through the theory of series of constants, focusing instead on series of functions. Extensive coverage is given for power series and Fourier series (expressing a periodic function as a sum of functions $\sin nx$ and $\cos nx$). This is a particularly good first reference on the latter, all the more valuable since introductory material on Fourier series not requiring advanced modern analysis can be difficult to find.
- Stanaitis, O. E., *An Introduction to Sequences, Series, and Improper Integrals*, Holden-Day, 1967

Books with Excellent Material on Infinite Series

- Apostol, T. M., *Mathematical Analysis, 2nd ed.*, Addison-Wesley (NY), 1974
- Gordon, Russell A., *Real Analysis: A First Course, 2nd ed.*, Addison-Wesley (NY), 2002
- Kosmala, Witold A. J., *A Friendly Introduction to Analysis: Single and Multivariable, 2nd ed.*, Prentice Hall (NJ), 2004
  - This textbook is very suitable for upper-level undergraduates and lower-level graduates. The "Friendly" in the title does well to describe the conversational, approachable tone that contextualizes the material of advanced real analysis very well, making this

book a suitable transition from basic calculus to higher mathematics. However, this does not come at the expense of mathematical rigor. The true-or-false questions accompanying each chapter are rewarding and well-chosen, and did a lot to inspire us to include our own such questions.

- Olmstead, John M. H., *Advanced Calculus*, Appleton-Century-Crofts (NY), 1961
  - A frequently used advanced calculus text used by students over a period of about three decades, this book has chapters (2, 11, 12, 13) on sequences and series. The exercise sets are quite appealing.
- Stewart, James, *Calculus: Early Transcendentals, 5th ed.*, Thomson, Brooks/Cole (CA), 2003
- Varberg, Dale, Edwin Purcell and Steven Rigdon, *Calculus, 8th ed.*, Prentice Hall (NJ), 2000
  - Two truly wonderful textbooks for undergraduates. All the standard calculus concepts are covered here thoroughly and in a very readable style. The examples and visuals are plentiful and well-done, and there are extensive exercise sections of varying difficulty. In every way these are good texts.

Sources for Excellent Problems Related to Infinite Series

- Biler, Piotr and Witkowski, Alfred, *Problems in Mathematical Analysis*, Marcel Dekker, Inc. (NY), 1990
- Kaczor, W. J. and Nowak, M. T., *Problems in Mathematical Analysis I: Real Numbers, Sequences and Series*, American Mathematical Society (RI), 2000
  - This book has a similar flavor to the Polya-Szego book (see below for annotation). The exercises here are generally easier and more accessible, and in some sense more enjoyable to many students. Here you will certainly want a selection of external analysis references to support the problem solver. Again, this is a book you should not fail to read if you are looking for problems, either for your own enjoyment and development or for a class.
- Larson, Loren C., *Problem-Solving Through Problem Solving*, Springer-Verlag (NY), 1983
- Polya, G. and Szego, G. *Problems and Theorems in Analysis, Vol. 1*, Springer-Verlag (NY), 1972
  - This book is intended to teach a full course of analysis by means of problem solving, and it works on that level for the brave student. The exercises are very challenging, but this book is not just a string of challenging exercises. On the contrary, organizational choices help the reader to apply recent previous problems to solve new ones, collections of problems combining into major theorems. Also, the solutions are uniformly well-written. As a pedagogical technique, this problem course may not be for everybody, but as a source for exercises, theorems, and related clusters of problems, one can hardly do better.

Pleasurable Reading

- Dunham, William, *The Calculus Gallery*, Princeton University Press, 2005
- Dunham, William, *Euler, the Master of Us All*, Mathematical Association of America (DC), 1999
- Dunham, William, *Journey Through Genius: The Great Theorems of Mathematics*, John Wiley and Sons, Inc. (NY), 1990
  - The three books by William Dunham are pedagogically excellent, well-written, and valuable resource books on a wide range of mathematical topics. Some of what he offers inspired one of the authors of this book to develop a deeper fondness for infinite series. For further information on infinite series, we suggest you open any of these books to the index, look under series, or infinite series, and read what is offered. Indeed, you may also find yourself wanting to read each book cover to cover. Dunham was awarded the 1993 George Polya Award of the MAA for excellence in expository writing about mathematics, which speaks much about his writings. They are truly a pleasurable read, an assessment that is shared not only by professional mathematicians but also by others who appreciate technical details presented correctly, with clarity, and in an engaging style.
- Havil, Julian, *Gamma: Exploring Euler's Constant*, Princeton University Press, 2003
  - *Gamma* by Havil is overall a very readable book in a similar spirit to the Dunham books. Chapters 2, 3, and 13 entitled respectively "The Harmonic Series", "Sub-Harmonic Series", and "It's a Harmonic World" fit nicely with the spirit and content of our book.

Journal Articles

MATHEMATICS MAGAZINE

Year	Vol./No.	Page(s)	Author	Title
1943	17:7	292–295	William E. Byrne	An Infinite Series
1948	22:1	53	H. A. Simons	Classroom Discussion of a Question on Infinite Series
1955	29:2	88	R. Lariviere	On a Convergence Test for Alternating Series
1967	40:3	120–128	B. J. Cerimele	Extensions on a Theme Concerning Conditionally Convergent Series
1971	44:5	273–276	Kenneth Williams	On $\sum_{n=1}^{\infty} \frac{1}{n^{2k}}$
1972	45:3	148–149	D. P. Giesy	Still Another Elementary Proof of $\sum_{k=1}^{\infty} \frac{1}{k^2} = \frac{\pi^2}{6}$
1973	46:1	40–41	Miltiades S. Demos	Class Notes on Series Related to the Harmonic Series
1974	47:4	197–202	E. L. Stark	The Series $\sum_{k=1}^{\infty} k^{-s}$, $s = 2, 3, 4, \ldots$, Once More
1978	51:2	83–90	Ralph P. Boas	Estimating Remainders
1978	51:4	235–237	Mark A. Pinsky	Averaging an Alternating Series
1979	52:3	178	William J. Knight	Convergence and Divergence of $\sum_{n=1}^{\infty} \frac{1}{n^p}$
1979	52:5	321	Paul Erdos & Robert Clark	Irrational Sum for a Series (Problem 1048)
1981	54:5	244–246	Richard Beigel	Rearranging Terms in Alternating Series
1983	56:4	232–235	Gerald Jungck	An Alternative to the Integral Test
1983	56:5	307–314	Morris Kline	Euler and Infinite Series
1984	57:4	215–216	Nobuo Adachi	An Infinite Series for Pi with Determinants
1984	57:4	228–229	Yaser S. Abu-Mostafa	A Differentiation Test for Absolute Convergence
1985	58:5	283–284	David Callan	Another Way to Discover that $\sum_{n=1}^{\infty} \frac{1\cdot 3\cdot 5\cdots(2n-1)}{n 2^{n+1} !} = \ln 2$
1986	59:3	176–178	L. Christophe, M. Finn & J.Crow	Two Series Involving Zeta Function Values
1987	60:5	282	G. Kimble	Euler's Other Proof
1990	63:5	291–306	Ranjan Roy	The Discovery of the Series for π by Leibniz, Gregory, and Nilakantha
1991	64:1	53–55	Courtney Moen	Infinite Series with Binomial Coefficients
1991	64:5	349	Dennis C. Russell	Another Eulerian-type Proof
1992	65:5	313–322	Jeffrey Nunemacher	On Computing Euler's Constant

MATHEMATICS MAGAZINE

Year	Vol./No.	Page(s)	Author	Title
1993	66:1	62–63	Howard Morris, Tom Apostol & Robert Wagner	A Maclaurin Series
1994	67:4	301–302	K. P. S. Bhaskara Rao	A New Test for Convergence of a Series
1995	68:4	289–293	G. H. Behforooz	Thinning Out the Harmonic Series
1995	68:4	298–301	Bruce Christianson	Condensing a Slowly Convergent Series
1996	69:2	122–125	R. A. Kortram	Simple Proofs for $\sum_{k=1}^{\infty} \frac{1}{k^2} = \frac{\pi^2}{6}$ and $\sin x = x \prod_{k=1}^{\infty} \left(1 - \frac{x^2}{k^2\pi^2}\right)$
1997	70:1	43-45	Gerald A. Heuer	More on Thinned-out Harmonic Series
1997	70:3	214–215	D. Cruz-Uribe	Relation between Root and Ratio Test
1998	71:1	42–50	Rick Kreminski	Differentiating Among Infinite Series
1998	71:4	313–314	Marc Frantz	The Telescoping Series in Perspective
1999	72:1	45–51	J.Efthimiou	Finding Exact Values for Infinite Series
1999	72:4	317–319	Neville Robbins	Revisiting an Old Favorite, $\zeta(2m)$
1999	72:5	347–355	Michael Goar	Olivier and Abel on Series Convergence: An Episode from Early 19th Century Analysis
2000	73:2	154–155	Ji Chunggang & Chen Yonggao	Euler's Formula for $\zeta(2k)$, Proved by Induction on k
2003	76:2	142	Sidney H. King	Math Bite: Convergence of p-series

THE AMERICAN MATHEMATICAL MONTHLY

Year	Vol./No.	Page(s)	Author(s) and Title
1906	13:5	97-100	O. L. Callecot, The Approximate Summation of n Terms of Any Harmonic Series
1907	14:12	215–217	Henry Kemmerling, Note on a Formula for the Summation of Certain Power Series
1914	21:2	48–50	A. J. Kempner, A Curious Convergent Series
1916	23:5	149–152	Frank Irwin, A Curious Convergent Series
1916	23:8	302–303	A. J. Kempner & E. J. Moulton, Solution to Problem 453
1918	25:4	186–204	Raymond W. Brink, A New Integral Test for the Convergence and Divergence of Infinite Series
1919	26:9	392–395	O. Schmiedel, On the Summation of Certain Series
1923	30:1	30–31	Louis Weisner, Note on the Summation of Certain Series
1928	35:8	433–435	Joseph Grant, The Summation of a Class of Series
1931	38:4	205–209	Raymond W. Brink, A Simplified Integral Test for the Convergence of Infinite Series
1932	39:2	62–71	C. N. Moore, Summability of Series
1933	40:4	216–218	W. F. Libby, A Convergence Test and a Remainder Theorem
1936	43:1	9–21	Irwin Roman, An Euler Summation Formula
1939	46:7	434–436	N. J. Lennes, The Ratio Test for Convergence of a Series
1939	46:6	349–352	A. W. Boldyreff, A Note on an Algebraic Identity
1939	46:6	338–341	Hukam Chand, On Some Generalizations of Cauchy's Condensation and Integral Test
1939	46:8	486–492	J. W. Bradshaw, Modified Series
1941	48:2	93–98	I. E. Perlin, Series with Deleted Terms
1941	48:3	180–185	C. T. Rajagopal, Remarks on Some Generalizations of Cauchy's Condensation and Integral Test
1942	49:7	441–446	T. H. Hildebrandt, Remarks on the Abel-Dini Theorem
1943	50:5	318–319	Richard Bellman, A Note on the Divergence of a Series
1945	52:2	70–72	R. W. Hamming, Convergent Monotone Series
1945	52:7	365–377	I. M. Sheffer, Convergence of Multiply-Infinite Series
1948	55:3	153–155	R. J. Duffin, A generalization of the Ratio Test for Series
1949	56:3	170–172	Morgan Ward, A Generalized Integral Test for Convergence of Series
1949	56:5	325–328	Robert Stalley, A Generalization of the Geometric Series

THE AMERICAN MATHEMATICAL MONTHLY

Year	Vol./No.	Page(s)	Author(s) and Title
1953	60:1	19–25	G. T. Williams, A New Method of Evaluating $\zeta(2n)$
1954	61:5	331–334	T. A. Newton, A Note on a Generalization of the Cauchy-Maclaurin Integral Test
1957	64:5	338–341	E. Baylis Shanks, Convergence of Series with Positive Terms
1962	69:3	215–217	Philip Calabrese, A Note on Alternating Series
1966	73:5	521–525	G. T. Cargo, Some Extensions of the Integral Test
1966	73:8	822–828	Paul Johnson & Ray Redheffer, Scrambled Series
1968	75:9	964-968	George Brauer, Series Whose Terms are Obtained by Iteration of a Function
1970	77:3	285–287	H. Miller, A Ratio-Type Convergence Test
1971	78:2	164–170	O. E. Stanaitis, Integral Tests for Infinite Series
1971	78:8	864–870	R. P. Boas Jr. & J. W. Wrench Jr., Partial Sums of the Harmonic Series
1972	79:6	634–635	G. J. Porter, An Alternative to the Integral Test for Infinite Series
1973	80:4	424–425	Ioannis Papadimitriou, A Simple Proof of the Formula $\sum_{n=1}^{\infty} \frac{1}{n^2} = \frac{\pi^2}{6}$
1973	80:4	425–431	Tom M. Apostol, Another Elementary Proof of Euler's Formula for $\zeta(2n)$
1974	81:10	1067–1086	Raymond Ayoub, Euler and the Zeta Function
1975	82:8	827–829	Mieczyslaw Altman, An Integral Test for Series and Generalized Contractions
1975	82:9	931–933	A. D. Wadhwa, An Interesting Subseries of the Harmonic Series
1977	84:4	237–258	R. P. Boas, Jr., Partial Sums of Infinite Series and How They Grow
1978	85:8	661–663	A. D. Wadhwa, Some Convergent Subseries of the Harmonic Series
1979	86:5	372–374	Robert Baillie, Sums of Reciprocals of Integers Missing a Given Digit
1979	86:8	679–681	J. R. Nurcombe, A Sequence of Convergence Tests
1981	88:1	33–40	Paul Schaefer, Sum-preserving Rearrangements of Infinite Series
1983	90:8	562–566	Bertram Ross, Serendipity in Mathematics (Summing a Specific Series)
1984	91:10	629–632	Murray Schechter, Summation of Divergent Series by Computer
1985	92:7	449–457	D. H. Lehmer, Interesting Series Involving the Central Binomial Coefficient
1987	94:7	662–663	Boo Rim Choe, An Elementary Proof of $\zeta(2) = \pi^2/6$
1992	99:7	622–640	J.M. Strange, Series and High Precision Fraud
1992	99:7	649–655	Bart Braden, Calculating Sums of Infinite Series

THE AMERICAN MATHEMATICAL MONTHLY

Year	Vol./No.	Page(s)	Author(s) and Title
1994	101:5	450–452	Jingcheng Tong, Kummer's Test Gives Characterizations for Convergence or Divergence of all Positive Series
1995	102:9	817–818	Hans Samelson, More on Kummer's Test
1998	105:6	552–554	Charles Kicey & Sudhir Goel, A Series for lnk
2001	108:9	851–855	Ákos László, The Sum of Some Convergent Series
2002	109:2	187–188	Liu Zheng, An Elementary Proof for Two Alternating Series
2002	109:2	196–200	Josef Hofbauer, A Simple Proof of $\zeta(2) = \pi^2/6$ and Related Identities
2003	110:7	561–573	Noam D. Elkies, On the Sums $\sum_{k=-\infty}^{\infty}(4k+1)^{-n}$
2003	110:6	540–541	James D. Harper, Another Simple Proof of $\zeta(2) = \pi^2/6$
2003	110:1	57	Uri Elias, Rearrangement of a Conditionally Convergent Series
2004	111:1	32–38	Steven Krantz & Jeffery McNeal, Creating More Convergent Series
2004	111:1	52–54	Thomas Osler, Find $\zeta(2p)$ from a Product Series
2004	111:5	430–431	Hirofumi Tsumura, An Elementary Proof of Euler's Formula for $\zeta(2m)$
2004	111:10	913–914	Jon R. Stefausson, Forward Shifts and Backward Shifts in a Rearrangement of a Conditionally Convergent Series

THE COLLEGE MATHEMATICS JOURNAL

Year	Vol./No.	Page(s)	Author	Title
1978	9:1	18–20	Andris Niedra	Geometric Series on the Gridiron
1978	9:1	46–47	Louise S. Grinstein	A Note on Infinite Series
1978	9:2	105–106	Peter A. Lindstrom	A Note on the Integral Test
1978	9:3	191	Thomas W. Shilgalis	Flow Chart for Infinite Series
1979	10:3	198–199	W. G. Leavitt	The Sum of the Reciprocals of the Primes
1982	13:3	199	Franklin Kemp	Infinite Series Flow Chart for the Sum of $a(n)$
1985	16:2	135–138	F. Brown, L. Cannon J. Elich & D. Wright	On Rearrangements of the Alternating Harmonic Series
1986	17:1	66–70	Paul Schaefer	Sums of Rearranged Series
1986	17:2	165–166	J. Richard Morris	Counterexamples to a Comparison Test for Alternating Series
1987	18:1	18–23	William Dunham	The Bernoullis and the Harmonic Series
1987	18:5	410	A. R. Amir-Moez	A Simple Proof of Series Convergence
1989	20:4	329–331	Alan Gorfin	Evaluating the Sum of the Series $\sum_{k=1}^{\infty} \frac{k^j}{M^k}$
1997	28:4	296–297	J. Marshall Ash	Neither a Worst Convergent Series nor a Best Divergent Series Exists
1997	28:3	209–210	Michael Ecker	Divergence of the Harmonic Series by Rearrangement
1997	28:5	368–376	Rick Kreminski	Using Simpson's Rule to Approximate Sums of Infinite Series
1999	30:1	34	Andrew Cusumano	Boxed Note (Harmonic Series Diverges)
1999	30:2	145–146	A. Kheyfits, et. al.	A Series Whose Sum is $\ln k$
2000	31:3	178–181	Frank Burk	Summing Series via Integrals
2001	32:2	119–122	James Lesko	A Series for $\ln k$
2001	32:3	201–203	David E. Kullman	What's Harmonic about the Harmonic Series
2001	32:3	206–208	Rasul Khan	Convergence-Divergence of p-Series
2001	32:5	377–380	Russell A. Gordon	Sum Rearrangements
2002	33:2	143–145	Leonard Gillman	The Alternating Harmonic Series
2002	33:2	168	Sidney Kung	An Application of Condensation
2002	33:3	233–234	Bernard August & Thomas J. Osler	Divergence of Series by Rearrangement
2002	33:4	314–316	Xianfu Wang	Convergence-Divergence of p-Series
2002	33:5	405–406	Fredrick Hartman & David Sprows	Investigating Possible Boundaries Between Convergence and Divergence
2004	35:1	34–39	Harlan Brothers	Improving the Convergence of Newton's Series Approximation for e
2004	35:1	43–47	Beata Randrianantoanina	A Visual Approach to Geometric Series
2004	35:3	171–182	James Lesko	Sums of Harmonic-Type Series
2004	35:3	183–191	C. Feist & R. Naimi	Almost Alternating Harmonic Series

Index

About the Authors

Daniel Donald Bonar was born in Murraysville, West Virginia on July 7, 1938, the son of Nelson Edward Bonar II and Ada Polk Bonar. He graduated from Ravenswood High School and was awarded a four-year full-tuition scholarship to West Virginia University where he received the BS in Chemical Engineering in 1960. While at WVU, he was a member of the physics, chemistry, and chemical engineering honoraries and served as President of Tau Beta Pi, the engineering honorary. Two National Science Foundation Fellowships supported his graduate work in mathematics. He received the MS from WVU in 1961 with a major in mathematics and a minor in physics. His PhD work was in complex analysis at Ohio State University where he graduated in 1968. In 1965 Don Bonar joined the faculty of Denison University in Granville, OH where he has been teaching mathematics, statistics, and computer science.

Awards received by Don Bonar include the Richard King Mellon Foundation Award for excellence in teaching and scholarship in 1973 and the Sears-Roebuck Teaching Excellence and Community Leadership Award in 1991. In 1995 he was appointed to the newly created George R. Stibitz Distinguished Professorship in Mathematics and Computer Science. In 1999 Don was inducted into the Academy of Chemical Engineers at West Virginia University. He is the author of a book entitled *On Annular Functions*, and is co-author on several research papers. He has published joint work with the internationally acclaimed Hungarian mathematician Paul Erdős. Community service includes membership on the Granville Foundation, the Granville Development Commission, the Licking County (OH) Joint Vacation School Board, and serving as President of the Granville Exempted Village School Board.

Don and his wife Martha Baker Bonar are the parents of Mary Martha, a student in medical school at Ohio University. The Bonars enjoy time at their farm, family-owned since 1869, in West Virginia.

Michael John Khoury, Jr. was born to Michael Sr. and Amy Khoury in Cuyahoga Falls, OH on April 28, 1981. He graduated as co-valedictorian from Brother Rice High School in Bloomfield Hills, MI. He studied for four years at Denison University in Granville, OH under the Wells Scholarship. He was a member of the mathematics and the education honoraries and president of the former; he also served as a departmental fellow in the Mathematics and Computer Science Department. Michael Khoury graduated in 2003 with a BS in Mathematics and Education, as well as the Presidential Medal (awarded to only six members of his class). He is currently pursuing a PhD in number theory at the Ohio State University with the support of the National Science Foundation.

Michael Khoury was a prize winner in the Putnam Mathematical Competition, and participated twice in the Math Olympiad Summer Program in Lincoln, NE, a four-week program open to the top performers on the USA Math Olympiad.